## Praise for *Bringing* Columbia *Home*

**Voted the Best Space Book of 2018 by the Space Hipsters**

"In *Bringing* Columbia *Home*, Mike Leinbach and Jonathan Ward have vividly captured the intensity of those very difficult days. They tell the story with compassion but without pulling any punches. The book also reminded me of the spirit of the American people who selflessly worked together to help NASA in its hour of greatest need. It's a message we all need to remember these days."

**—Scott Kelly**

"*Bringing* Columbia *Home* explains a disaster in the Texas skies—and how thousands on the ground helped. . . . [It] shines brightest in telling the story of the search-and-recovery effort."

***—Dallas News***

"Riveting."

**—*Air & Space* magazine (Smithsonian)**

"A grimly captivating new history of the loss of the space shuttle *Columbia*. . . . Leinbach and Ward set their account apart from other '*Columbia*' books by following the story from its central tragedy to its almost unthinkably sad immediate aftermath. . . . Despite the dramatic tragedy at the beginning of the book, it's the quiet stories of perseverance and camaraderie [in the recovery effort] that will linger longest with the reader."

***—Christian Science Monitor***

"How glowing is our praise of this book? It simply cannot be higher. This book needs to be required reading in high schools and colleges across the United States."

***—Spaceflight Insider***

"*Bringing* Columbia *Home* is a compelling, personal story about the Columbia accident and the efforts to recover—both the debris from the shuttle, and from the accident itself. It's a reminder that, as we look at the big-picture policy perspective of human spaceflight, it's also a very personal matter for those who put their lives on the line to fly, and those who support them."

—***Space Review***

"A gripping account of a fatal tragedy and the impressive and deeply emotional human response that ensued."

—***Kirkus Reviews*, ★starred review★**

"Gripping and dramatic . . . It's an important and fascinating chapter in space history, and it finally gets the full treatment it deserves. As told by someone who was involved in the effort from the beginning, it's also a deeply personal and moving story."

—***Booklist***

"Fast-paced and affecting . . . It is a moving and sometimes uncomfortably close account. . . . The unadorned, multisensory narration richly depicts the emotions and everyday acts of heroism of all involved."

—***Publishers Weekly***

"The book *Bringing* Columbia *Home* presents vivid details of the preparation and the aftermath of that fateful day when *Columbia* exploded. I am so grateful that the heartwarming story of the people of East Texas rallying to help the grim search has been brought forth by Michael Leinbach and Jonathan Ward. It is a remarkable account of what a team of professionals with an untrained but willing army of volunteers could achieve."

**—The Honorable Kay Bailey Hutchison, Senator for Texas, 1993-2013**

"Mike and Jonathan have done a brilliant job capturing the depth of emotion and human engagement of what has been covered by others only as a technical investigative treatment. In doing so, they have made the story very personal for the thousands of people who invested themselves in this critical chapter of space exploration history. This is a valuable contribution about a defining moment that demonstrates NASA's resolve and the selfless generosity of the American spirit."

**—Sean O'Keefe, former administrator of NASA**

"Mike and Jonathan have written an important book about the greatness of the United States and the American people in responding to a national tragedy. This book brought back many memories—and some tears—as I recalled the selfless cooperation of countless agencies and the outpouring of support and prayers from the nation's citizens, all aimed at getting NASA and the Space Shuttle flying again."

**—Jerry L. Ross, former astronaut, retired USAF Colonel, and author of *Spacewalker: My Journey in Space and Faith as NASA's Record-Setting Frequent Flyer***

"Spaceflight is an inherently risky business. I had more than my share of close calls in my career. But the Apollo 1 fire and the *Challenger* and *Columbia* accidents were grim reminders that we sometimes have to pay a very dear price in the cause for human advancement. In *Bringing* Columbia *Home*, Mike Leinbach and Jonathan Ward tell the remarkable story of what NASA and the American people did supremely well after a crisis: supported each other through difficult times, tirelessly looked for solutions, and then moved forward to accomplish bold goals. This engaging and inspiring book reminds us of what Americans look like at their best—cooperative, compassionate, and committed."

**—James Lovell, former astronaut and coauthor of *Apollo 13***

"I was privileged to call *Columbia* my home in space for eighteen days. Thanks to this moving and heartfelt story, now I know how many thousands gave their all to bring this storied ship and her crew to an honored rest."

**—Tom Jones, former astronaut and author of *Ask the Astronaut* and *Sky Walking: An Astronaut's Memoir***

"*Bringing* Columbia *Home* is about tragedy and how tragedy is overcome. Leinbach and Ward have written an intensely compelling book with life lessons for everyone in the space community and ordinary life. It is an intensely human and technological drama."

**—Hugh Harris, former director of public affairs, NASA Kennedy Space Center**

"*Bringing* Columbia *Home* is a wonderful contribution to spaceflight history, a previously untold story of heroism related responsibly, compassionately, and accurately. The depiction of the entire NASA family's participation in the recovery and investigation is touching and respectful. People are going to love this book."

**—Susan Roy, author of *Bomboozled: How the US Government Misled Itself and Its People into Believing They Could Survive a Nuclear Attack***

Bringing *Columbia* Home

# Bringing *Columbia* Home

## The Untold Story of a Lost Space Shuttle and Her Crew

Michael D. Leinbach
and Jonathan H. Ward

Foreword by Astronaut Robert Crippen
Epilogue by Astronaut Eileen Collins

With a New Preface by the Authors

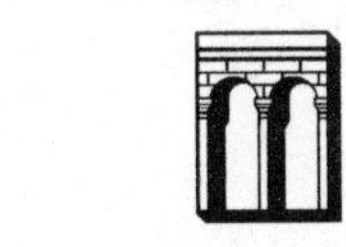

Arcade Publishing • New York

First paperback edition

Arcade Publishing books may be purchased in bulk at special discounts for sales promotion, corporate gifts, fund-raising, or educational purposes. Special editions can also be created to specifications. For details, contact the Special Sales Department, Arcade Publishing, 307 West 36th Street, 11th Floor, New York, NY 10018 or arcade@skyhorsepublishing.com.

Arcade Publishing® is a registered trademark of Skyhorse Publishing, Inc.®, a Delaware corporation.

Visit our website at www.arcadepub.com.
Visit the authors' website at http://www.bringingcolumbiahome.com.

10 9 8 7 6 5 4

Library of Congress Cataloging-in-Publication Data

Names: Leinbach, Michael D., author. | Ward, Jonathan H., author.
Title: Bringing Columbia home : the final mission of a lost space shuttle and her crew / Michael D. Leinbach, former launch director, John F. Kennedy Space Center, NASA, and Jonathan H. Ward ; foreword by astronaut Robert Crippen ; epilogue by astronaut Eileen Collins.
Description: First edition. | New York : Arcade Publishing, [2017] | Includes bibliographical references and index.
Identifiers: LCCN 2017046190 (print) | LCCN 2017050797 (ebook) | ISBN 9781628728521 (ebook) | ISBN 9781628728514 (hardcover : alk. paper) | ISBN 9781948924610 (paperback)
Subjects: LCSH: Columbia (Spacecraft)–Accidents. | Space Shuttles–Accidents–United States. | Space vehicles–Recovery. | Space vehicle accidents–United States. | Astronauts–Accidents–United States.
Classification: LCC TL867 (ebook) | LCC TL867 .L45 2017 (print) | DDC 363.12/4–dc23
LC record available at https://lccn.loc.gov/2017046190

Cover design by Erin Seaward-Hiatt
Cover photo credit: AP

Printed in the United States of America

**To *Columbia*'s final crew**
Rick, Willie, Laurel, Mike, KC, Dave, and Ilan

**To Buzz Mier and Charles Krenek**
who gave their lives in the search for *Columbia*

**And to the 25,000 heroes**
who brought *Columbia* and her crew home one last time

# CONTENTS

# FOREWORD

The story of the recovery of the space shuttle *Columbia* deserves to be better known and celebrated.

On February 1, 2003, I was at home enjoying my retirement, when I received a phone call from my daughter Susan, who worked on the Space Shuttle Program at Johnson Space Center. She said, "Dad, they've lost contact with *Columbia*."

Her words felt like a heavy punch in my gut. *Columbia* was on reentry coming home from her STS-107 mission. Susan and I both knew that if they had lost contact during reentry, then the vehicle was lost. Only three days earlier, we had marked the seventeenth anniversary of the loss of *Challenger*.

The Space Shuttle Program consumed a major portion of my life. I worked on the shuttle during its development, and I was fortunate enough to be selected to fly with John Young on *Columbia* during its maiden flight. That first flight was more successful than any of us involved with the program could have hoped for. This was the first time astronauts launched on a vehicle that had not first been tested in an unmanned flight. It was the first crewed vehicle to use solid rocket boosters, and it was the first spacecraft to return to a landing on a runway. John and I were very proud of *Columbia*'s performance on that initial test flight.

I flew three more times after my first mission, and I had been preparing to command the initial shuttle flight out of Vandenberg Air

Force Base when the *Challenger* was lost. I became deeply involved with the accident investigation and eventually moved into Shuttle Program management. That move was prompted by my desire to get shuttles flying again, safely. We implemented many changes to the program, including to the hardware, software, and management. I believed those changes and careful oversight would ensure that we wouldn't lose another vehicle.

Then we lost *Columbia*. Not only was it a vehicle that I was very fond of, but I also knew that the program would probably not survive losing a second shuttle and crew. The subsequent investigation found the physical cause of the accident, and also showed that NASA had forgotten some of the lessons learned from the *Challenger* loss. Safety and management practices had eroded over time. My concern over the cancellation of the program proved valid. NASA decided to return the shuttles to flight only to complete the International Space Station, but then the orbiters would be retired. An era in human spaceflight had come to an end.

*Columbia*, being the first shuttle built, weighed more than her sister ships. After building *Columbia*, NASA determined that an orbiter's aft structure did not have to be as beefy. Consequently, *Columbia* didn't get some of the more sexy assignments due to her lower performance capability compared to the other orbiters. Still, she flew all her missions exceptionally well. She was a proud old bird. I know she did her best to bring her last crew home safely, just as she had done twenty-seven times before. However, her mortal wound was just too great.

I was not the only one who felt close to *Columbia*. All the women and men on the ground who prepared and flew her missions felt that same connection. She wasn't just an inanimate machine to them. Their shock at her loss was as deep as mine, or deeper. I knew how the people felt from my years at Johnson Space Center as an astronaut and then my time as director of Kennedy Space Center. *Columbia*'s loss was intensely personal to everyone involved.

Those dedicated workers now had a compelling desire to determine the cause of the accident and to return the other vehicles to flight status. That involved finding the remains of the crew and as much of the vehicle debris as possible. Debris retrieval is essential in any accident investigation. We needed the wreckage to determine what had happened to *Columbia*.

My experience with *Challenger* told me this was going to be a long and tough task. *Challenger*'s debris was in a relatively tight cluster but was submerged because the accident occurred on ascent, just off the Florida coast. Since *Columbia*'s accident occurred on reentry, her debris was spread over a large area in East Texas and Louisiana.

The NASA team, mostly from Kennedy and Johnson, set about the task of finding *Columbia* with the same diligence that they had for *Challenger*. Their job was a tough one, physically and emotionally. This book, *Bringing* Columbia *Home*, demonstrates the dedication of the women and men who undertook this extremely trying job at a time when their hearts were full of sorrow. With the help of many people and agencies, they recovered the crew remains and a remarkable portion of the debris. NASA took that debris home to Kennedy, where it provided the physical evidence we were hoping to find. That hardware debris, along with the telemetry data from the vehicle during reentry, conclusively proved the cause of the accident. That enabled the NASA team to correct the problem and return the shuttle to flight.

Even with the loss of two vehicles and fourteen wonderful people, I am still proud of the Shuttle Program's legacy. It was a space vehicle like no other, with the capability to lift very large payloads into space along with crews. The ability to put crews and payloads together proved extremely valuable. People have questioned the combining of the two, but I think that helped make the machine the magnificent vehicle it was. Early on, it carried out some very important Department of Defense missions that played a significant role in the Cold War. The shuttle allowed us to revolutionize our knowledge of our solar system with missions like *Magellan*, *Galileo*, and *Ulysses*. It also drove us to rewrite the books on our knowledge of the universe with the great observatory missions such as Hubble, Compton, and Chandra.

Especially with the repair missions to the Hubble Space Telescope, the shuttle demonstrated the benefits of combining crew and payload on the same vehicle. Lofting humans and payloads together on missions also allowed us to construct the greatest engineering marvel of all time, the International Space Station.

It will be a long time from now, if ever, that we see another vehicle with such an astounding capability.

And there will never be another bird like *Columbia.*

**Capt. Robert L. Crippen, USN, Retired**
Pilot, STS-1
Commander, STS-7, STS-41C, STS-41G

# PREFACE TO THE PAPERBACK EDITION

America was put to the test on the morning of February 1, 2003.

At Kennedy Space Center, we waited in silence for a space shuttle that would never make it home. At the same time, residents of an area stretching from Dallas to Fort Polk, Louisiana heard thousands of sonic booms followed by tens of thousands of pieces of the broken *Columbia* falling from the sky all around them.

What ensued was one of the most remarkable examples of civic response to tragedy that this country has ever seen. Americans put aside their political differences. For a brief period, we stopped arguing over the impending Iraq war. Instead, we pulled together to find *Columbia*'s crew and to solve the mystery of what doomed the ship.

This isn't a space book. It's a story of everyday people who found themselves at the center of an accident of unprecedented proportions and then rose to the challenge to help their country. Volunteers fought their way through briar-choked bayous of the Texas piney woods. Housewives cooked food for thousands of people who descended on a tiny town to help with the recovery. Schoolchildren made sandwiches and wrote notes of encouragement to the searchers. Astronauts combatted grief to look for their lost colleagues. Native American firefighters flew in from the West to comb every square foot of an area the size of Delaware. Technicians reassembled the broken pieces of their beloved shuttle to try to find the cause of the accident. Everyone played a part in what collectively became the largest ground search and recovery operation in US history.

Any one of them could easily have elected *not* to answer the call, for reasons that everyone would have understood. They chose instead to follow a path of hardship and uncertainty that ultimately led to a bittersweet victory for *Columbia*, her crew, and our country.

The humble residents of East Texas, and all the others who came to help, demonstrated what it means to serve without hesitation and without even being asked. As one resident told us, "I don't even want a 'thank you.' I did what I could. I did what I felt was right."

They were all heroes.

We had the good fortune to share our newly-released book with the people of Hemphill, Texas—"ground zero"—on the fifteenth anniversary of the accident. It was a profoundly moving experience to be at that hallowed ground and to thank so many of the people whose lives were forever changed when *Columbia* fell to Earth. Perhaps most gratifying of all was speaking with the spouses of several of *Columbia*'s crewmembers, who thanked us for telling the story of their terrible loss with dignity and respect. We've given talks to thousands of people about the book, but that day in East Texas will forever shine bright in our memories.

It has been an honor and privilege to tell the world the final story of America's first space shuttle. The story reminds us all how great we can be—and how remarkable we truly are—when tested in ways no one could possibly prepare for.

*Be of good courage . . .*

MIKE LEINBACH AND JONATHAN WARD

# PART I

# PARALLEL CONFUSION

We have seen this same phenomenon on several other flights and there is absolutely no concern for entry.

—*Email from Mission Control to* Columbia*'s crew,*
*January 23, 2003*

# Chapter I

# SILENCE AND SHOCK

**Kennedy Space Center**
**February 1, 2003**
**Mike Leinbach, Launch Director**

Twin sonic booms in rapid succession—one from the space shuttle's nose and one from its vertical tail—were always the fanfare announcing the arrival of the majestic winged spacecraft. Three minutes and fifteen seconds before landing, as the shuttle glided toward Kennedy Space Center (KSC), the loud and unmistakable double concussion would be heard up and down Florida's Space Coast. These booms would be our cue to start scanning the skies for our returning spacecraft, descending toward us at high speed in the distance.

*Columbia* and her crew of seven astronauts were coming home from sixteen days in orbit. After six million miles circling the Earth, they had reentered the atmosphere over the Pacific Ocean, crossed the California coast, and then flown over the Desert Southwest and Texas en route to Florida. These last few miles would be her victory lap in front of her astronaut crew members' families and the KSC personnel who tended her on the ground.

As KSC's launch director, I was one of the officials who would welcome *Columbia* home when she landed at 9:16 this cool morning. At 9:12, we listened and waited for the thunderous sonic booms, which would sound like the percussion of an artillery volley. But today the heavens were strangely silent.

Over the loudspeaker feed from Mission Control, we heard repeated calls to the crew: "*Columbia,* Houston. Comm check." Long moments of silence punctuated each call. "*Columbia*, Houston. UHF comm check."

I was confused and alarmed. I looked up at the clouds and turned to Wayne Hale, NASA's former ascent and entry flight director, and asked him, "What do you think?"

He thought for a moment and responded with a single word: "Beacons."

That one word hit me hard. The astronauts' orange launch and entry suits were equipped with radio beacons in case the crew needed to bail out during a landing approach. Hale clearly knew the crew was in trouble. He was already thinking about how to find them.

*My God.*

The landing countdown clock positioned between the runway and us ticked down to zero. Then it began counting up. It always did this after shuttle landings, but we had never really paid attention to it because there had always been a vehicle on the runway and that clock had become irrelevant.

*The shuttle is never late. It simply cannot be.*

*Columbia* wasn't here. She could not have landed elsewhere along the route. She was somewhere between orbit and KSC, but we didn't know where.

I tried to sort out my thoughts. Something was horribly wrong. An indescribably empty feeling swept over me. My position as launch director was one of knowledge and control. Now I had neither.

Kennedy Space Center and Cape Canaveral have seen more than their share of disasters. A launch catastrophe is unmistakable—tremendous noise, a horrendous fireball, and smoking debris falling into the ocean. My mind flashed back to the frigid morning of January 28, 1986. I had been standing outside and seen *Challenger* lift off from pad 39B, only to disappear into a violent conflagration shortly afterward.

I remember expecting—hoping—that *Challenger* would emerge from the fireball, fly around, and land behind me at the Shuttle Landing Facility. But we never saw *Challenger* again. I recalled leaving the site with a few friends as debris and smoke trails continued to rain down into the Atlantic, just off the coast. It was a terrible thing to witness.

This situation was completely different. Our emergency plans assumed that a landing problem would happen within sight of the runway, where a failed landing attempt would be immediately obvious to everyone. Today, there was nothing to see, nothing to hear. We had no idea what to do.

*Columbia* simply wasn't here.

We all knew something awful must have happened to *Columbia*, but our senses could tell us nothing. The audio feed from Mission Control had gone eerily silent.

The breeze picked up. Low rippling clouds masked the sun. The quiet was broken only by a few cell phones that began ringing in the bleachers where spectators and the crew's families were waiting. The astronauts in the ground support crew huddled briefly by the convoy command vehicle. Then they moved with quick determination toward the family viewing stand.

I glanced over at Sean O'Keefe, NASA's administrator. He appeared to be in shock. O'Keefe's associate administrator, former astronaut Bill Readdy, stood at his side. Readdy looked me in the eye and asked, "Contingency?" Unable to speak, I simply nodded.

Readdy carried a notebook containing NASA's agency-wide contingency plan for spaceflight emergencies. Ever the pragmatist, O'Keefe had ordered this plan updated within hours of becoming administrator in late 2001. Now, barely one year later, the plan had to be activated. The procedures designated Readdy to make the official call. He told O'Keefe that he was declaring a spaceflight contingency.

Gathering my thoughts and trying to keep my emotions in check, I told the officials to meet me in my office back at the Launch Control Center (LCC), about two and a half miles to the south. We could confer there in private and get more information about the situation.

KSC security personnel and astronaut escorts quickly led the crew's families away from viewing stands to a bus that would take them to the privacy of the crew quarters. The other spectators—many of whom were friends of the crew or members of the crew's extended families—were also ushered to waiting buses.

There was no announcement of what had happened, but everyone knew that it must be something dreadful. Few words were spoken. People wept and hugged one another as their initial emptiness slowly filled with grief.

In the utterly inadequate jargon of astronauts and space workers, this was going to be a *bad day*.

As I hustled back to my vehicle, I had no idea how this horrible day would unfold—or how inspiring its aftermath would ultimately be.

Chapter 2

# GOOD THINGS COME TO PEOPLE WHO WAIT

I began my twenty-seven years with NASA in 1984 as a structural engineer. With an undergraduate degree in architectural design and a master's in structural engineering—both from the University of Virginia—I was living my childhood dream! I couldn't believe I was working at Florida's Kennedy Space Center, designing portions of the launchpad platforms, emergency escape systems, and the like. I moved into an operations role shortly after the *Challenger* accident, becoming a NASA test director and a member of the shuttle launch team. I moved up fairly quickly and eventually was leading the launch countdown.

After a two-year stint as deputy director of the International Space Station Program Office at KSC, I became the eighth launch director of the Space Shuttle Program in August 2000. I was now responsible for all shuttle launch operations, including giving the final "Go"—or often "No-go"—for launch.

Through it all, I never lost touch with friends made along the way, nor did they stop reaching out to me. I like to think I was just a regular KSC guy who got a big job.

Some of my predecessors had a sort of old-school management style that entailed demanding action and acting aloof toward junior personnel. However, my openness, combined with coming up through the ranks as I did, earned me the moniker "the people's launch director." I was no overbearing type—but I was also no pushover. People always knew where they stood with me and what I expected of them. I publicly recognized superior performance, and also did some course correcting when necessary. It was a combination that worked well for the team and me. Together, we accomplished amazing things.

And what a thrill it was to work with the space shuttle! It was a masterpiece of American technological prowess—the pinnacle of NASA's manned spacecraft evolution. Each of the winged vehicles of the "Space Transportation System," which we called the "orbiter" or just simply "the shuttle," took off like a rocket from KSC and landed like a glider. Crews of up to seven astronauts[1] could work in a spacious shirtsleeves environment for missions lasting as long as sixteen days, while the temperature in the vacuum of space just outside their windows ranged from 250°F in direct sunlight to minus 250°F in the shade. They could also venture outside through an air lock to perform space walks. The shuttle's cargo bay carried payloads as large as a school bus.

Each orbiter—the size of a small commercial airliner—was lofted into Earth orbit bolted to an enormous external fuel tank and a pair of the most powerful solid propellant rocket boosters ever developed. The two solid rocket boosters turned 2.2 million pounds of fuel into energy—and speed—in the course of 127 seconds. The foam-covered fuel tank held about 1.6 million pounds of liquid oxygen and liquid hydrogen, which the shuttle's three main engines gulped dry in the space of eight and one half minutes, by which time the shuttle was in orbit and traveling 17,500 mph. Everything in the system except for the external tank could be reused.

NASA's shuttle fleet—*Columbia*, *Challenger*, *Discovery*, *Atlantis*, and *Endeavour*—flew 135 space missions between 1981 and 2011, carrying a total of 833 crew members.[2] Space shuttles took 3.5 million pounds of cargo into orbit during the Program. This included scores of different payloads—satellites, laboratories, planetary probes, NASA's Great Observatories (such as the Hubble Space Telescope), experiments, and space station modules.

There had never been such an amazing flying machine.

It was also far from perfect.

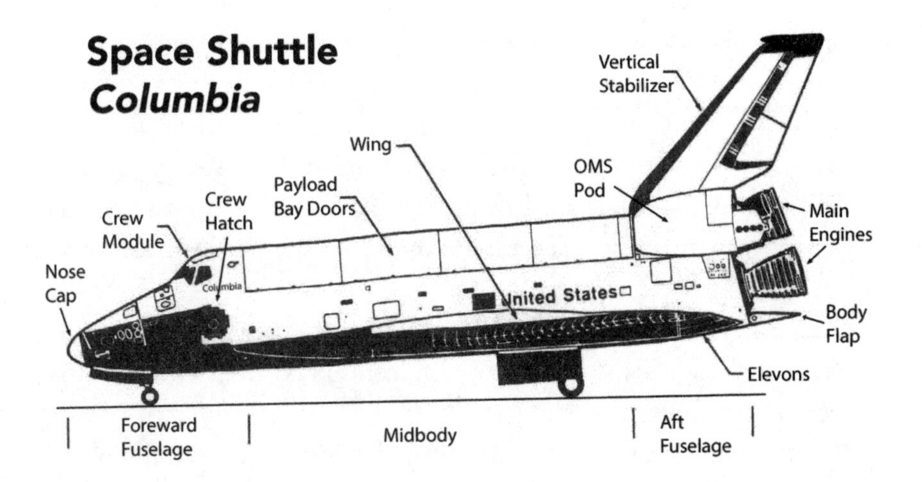


In 1997, NASA announced plans for a sixteen-day research mission, STS-107.[3] The new Spacehab double module, about the size of a school bus, would fly in the payload bay of *Columbia*, the flagship of the shuttle fleet. Spacehab was an orbital laboratory boasting a wide array of science and medical experiments, studying subjects as diverse from how various systems in the human body respond to weightlessness to how to grow protein crystals for cancer therapies. Spacehab was pressurized and connected to the shuttle's cockpit by a tunnel, allowing the astronauts to operate the research equipment in a shirtsleeves environment. NASA announced the mission's crew in July 2000.

United States Air Force Colonel Rick Husband was the mission commander and the man at the shuttle's controls. He had served previously as pilot—the second-in-command, who does not actually fly the shuttle—on STS-96. He was one of very few astronauts to be given command of a mission after only one previous spaceflight. A deeply religious man, Husband was renowned for his sense of humor, ability to build cohesive teams, and beautiful singing voice.

Commander William "Willie" McCool, the mission's pilot, was a US Navy test pilot and was on his first shuttle mission. His colleague Laurel Clark described him as a "ten-year-old trapped in the body of an eight-year-old" because of his boyish looks and youthful exuberance.[4]

Lieutenant Colonel Michael Anderson, USAF, served as payload commander for STS-107. He was a veteran of one previous shuttle mission and was the ninth African American to fly in space.

Kalpana "KC" Chawla, PhD, an aerospace engineer, was the first Indian-born woman in space. Flying on her second space mission, she was STS-107's flight engineer.

Captain Dave Brown, MD, a naval aviator and naval flight surgeon, was a mission specialist on his first spaceflight. Brown was the only unmarried member of the crew.

Commander Laurel Clark, MD was, like Brown, a naval flight surgeon and a mission specialist on her first spaceflight.

Colonel Ilan Ramon, a fighter pilot in the Israeli Air Force, was the specialist operating an experiment to observe dust storms in the Mediterranean and Israel. He was Israel's first astronaut, and this was his first space mission.

Between their selection and their flight, the crew spent more than 4,800 hours training for the mission and an additional 3,500 hours training to run the medical and scientific experiments in the Spacehab module. The many mission delays—thirteen in all, due to priority changes and hardware issues—enabled the crew to bond closely with one another. They spent nine nights camping in Wyoming as part of an outdoor leadership course in 2001. Brown carried a video camera everywhere to record the crew's preparations and commemorate their friendship.

A mission's commander sets the tone for how the crew interacts with the support teams on the ground. Some commanders were type A personalities—all business. Husband, on the other hand, was one of the warmest and most caring commanders imaginable. He, and by extension his crew, made everyone he worked with on the ground support teams feel like part of a family.

Robert Hanley, from Houston's Johnson Space Center (JSC), served as the interface between the astronaut crew and the teams at Kennedy who were preparing *Columbia* for her mission. Hanley got to know the STS-107 crew and their families intimately during the two years leading up to the mission. He said, "Hands down, 107 was the best crew I ever had. They were just awesome individuals. Rick set the stage that 'Hey, we're gonna be a warm, happy, fun crew,' and they were."

Ann Micklos, the lead airframe engineer for *Columbia* at KSC, was responsible for structural issues and the thermal protection system on the orbiter. Apart from her official role working with the shuttle, she had a unique relationship with the crew—she and Dave Brown had been dating since before his assignment to STS-107. Their connection further strengthened the personal relationship between the ground crews and *Columbia*. Ann said, "It wasn't just personal for me. It was personal for everyone who was working on that vehicle—they all knew me and knew I was dating Dave."

In June 2002—just prior to *Columbia*'s originally scheduled July launch date—Ann received a birthday package from Dave while visiting family in Connecticut. Inside was an empty watch box. Fearing that the watch had been stolen, she looked more carefully and found a note taped to the lid, which read, HELP! I'M BEING HELD HOSTAGE ABOARD THE SPACE SHUTTLE! Ann was ecstatic to hear that Dave would be flying with her watch on the mission.

A few months later, Ann and Dave ended their romantic relationship, but they remained very close friends. She participated as Dave's "stand-in spouse" at all of the traditional prelaunch activities attended by the crew's spouses and families.

—

*Columbia*, like her sister shuttles, lived at Kennedy Space Center when she was not in orbit. An orbiter might fly three times in a year, for an annual total of five to six weeks in orbit. The rest of the time, our ground teams at Kennedy took care of it.

Our people knew the actual flight hardware better than anyone—the whine of every cabin fan, the condition of every tile on the orbiter's belly, the twists and turns of the fuel piping in the engine compartment—and hundreds of people at KSC lived with the orbiter every day for much of their careers. The orbiter only left our care during the ten to fourteen days it was in flight.

As a reusable vehicle, the shuttle had to be inspected, repaired, and maintained after each flight. Its complex systems meant that this was no easy task. No matter how well a shuttle performed on its mission and how good it looked after a flight, preparing it for its next mission was never as simple as giving it a quick once-over. Our workers spent tens of thousands of man-hours checking and maintaining every system, replacing damaged thermal insulation tiles, reconfiguring the crew compartment and payload bay for the requirements of the next mission, changing the tires, and performing myriad other tasks to ensure the shuttle continued to meet its incredibly stringent reliability and safety requirements.

All this meant that the hands-on workers at Kennedy—primarily the engineers and technicians of our main contractors United Space Alliance (USA) and Boeing—were intimately familiar with each nut and bolt, wiring harness, coolant pipe, and every single one of the hundreds of thousands of parts on board the orbiter.

*Columbia* was a little different from her sister orbiters. As the first shuttle constructed for spaceflight, her structure and internal plumbing were unique. She had a different tile pattern and air lock, and she carried instrumentation that the other orbiters lacked. She was eight thousand pounds heavier than her sister ships. The differences were subtle, but they were significant enough that technicians who serviced the other three orbiters sometimes became frustrated if they were called over to work on *Columbia*. She developed a reputation at Kennedy for being the beloved black sheep of the fleet.

Rather than apologizing for her, the dedicated *Columbia* processing teams rallied around "their" ship and became even more close-knit as a consequence. They loved *Columbia* and her quirks. Many people specifically requested to work on *Columbia* because of her status as the flagship of the fleet.

Quality inspector Pat Adkins said: "You can't actually put into words exactly how you feel about a spacecraft. You use it, you learn it—you know where all its little idiosyncrasies and scars are. You know its weak spots, its strong points. They were all different. If you talk about a mission and don't talk about the spacecraft like an eighth member of the crew, it's like trying to tell the story of *Star Trek* without the *Enterprise*."

We all knew how he felt. *Columbia* was just as "alive" to us as the people who flew her.

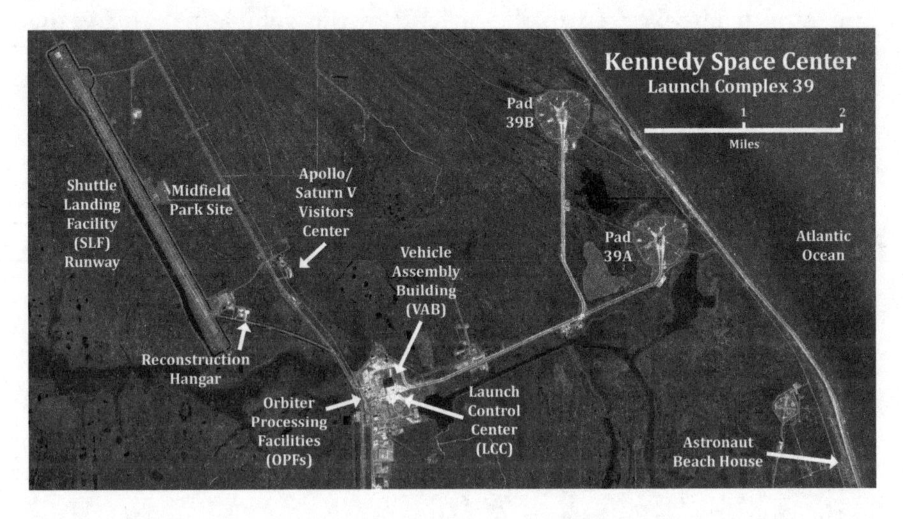


Astronauts typically spent most of their time in Houston training for their upcoming mission. Unlike the Apollo and earlier missions, where each space capsule only flew once, shuttle crews did not have a spacecraft that was uniquely "theirs." They could not work with their assigned vehicle until it returned from its latest mission. There were also no training simulators at KSC. The commander and pilot occasionally came to town to practice landing approaches in the Shuttle Training Aircraft at the KSC runway, but most of the crew usually did not visit Kennedy until their mission drew near.

STS-107 was an exception in that the facility where the Spacehab module was being prepared for the mission was located outside the southernmost security gate on the air force side of the property occupied by NASA and the air force. Marty McLellan, Spacehab's vice president of operations, set aside a desk for Rick Husband decorated with a HOME SWEET HOME plaque, because the crew was in town so frequently to train with the equipment.

While the astronauts were training, the shuttle was being prepared in an Orbiter Processing Facility (OPF)—one of three hangars adjacent to KSC's Vehicle Assembly Building (VAB).[5] The shuttles spent more time in the OPF than anywhere else. In those special hangars, some teams worked in the aft end of the vehicle to replace the engines and service the propulsion systems. Other teams worked to reconfigure the cargo bay and the crew compartment for the requirements of the mission.

Part of preparing for a mission included a "crew equipment interface test"—more of a weekend-long activity than an actual "test." The usual processing activities in the OPF were shut down and distractions were minimized, so the astronauts could spend time with the payload and the orbiter to get a feel for the configuration of their vehicle. Practicing in the simulators and the mock-ups at Houston was no substitute for the crew putting their hands on the actual flight hardware and seeing where everything was going to be stowed in the ship.

Pat Adkins remembered *Columbia*'s crew arriving with happy confidence on June 8, 2002. He said, "They were all smiles, especially Willie McCool. His was the biggest! As they passed by me, I looked them in the eyes and promised, 'We'll give you a good ride!'"[6]

The crew inspected the orbiter and checked out everything with which they would be working in orbit. The astronauts noted which cables were routed to which equipment items and looked behind panels and under the mid-deck floor. The crew requested that Velcro strips or stickers be put where they wanted them in the cabin. These strips would anchor cue cards, timers, and other items once the shuttle and her crew were weightless. It was the first time that many of the KSC ground support team worked directly with the astronauts for the mission. At the end of the activity, the astronauts and the ground workers posed for a picture together. It was an especially exciting moment for our processing team. Afterward, the astronauts and several of the KSC workers gathered at a local restaurant for food, drinks, and fellowship.

STS-107 was scheduled to fly in July 2002, but cracks discovered in the flowliners of *Atlantis*'s fuel system caused the whole fleet to stand down for inspections during the summer. Once the shuttles were cleared to fly, *Atlantis* took off with STS-112 in October 2002, and *Endeavour* rolled out to the launchpad later that month for STS-113. Then—finally!—it was time to roll *Columbia* over to the cavernous Vehicle Assembly Building and into the 525-foot-tall High Bay 1 for "stacking." *Columbia* was hoisted to a vertical orientation and mated to its external tank and solid rocket boosters on Wednesday, November 20, 2002.

*Columbia* rolled out to launchpad 39A on Monday, December 9. She had not even left on her mission, but engineers were already discussing plans for how to refit her with a new air lock once she returned. They needed her to fly one support mission to the International Space Station (ISS) if NASA was going to meet the Congressionally committed assembly schedule.[7]

—

Once *Columbia* was at the launchpad, the flight crew returned for a training session the week of December 16, which culminated in the terminal count demonstration test (TCDT). I greeted the crew with my traditional, "Welcome to TCDT Week!" at the Shuttle Landing Facility runway after they flew in from Houston in their T-38 jets. This was often the first time I had the opportunity to meet the rookie astronauts on a crew. I wanted the astronauts to feel comfortable with me—the man responsible for their safety on launch day.

TCDT week was full of activities to help the astronauts practice for a launch and to familiarize them with the systems that would save their lives if anything went wrong. The crew donned the orange pressure suits they would wear for launch and landing. They practiced emergency evacuation from the shuttle, running across the swing arm on the launch tower to the slidewire baskets that would take them to the perimeter of the launchpad. There, they would enter an underground concrete bunker and await instructions from the control room. Positioned adjacent to the bunker was an M-113 armored personnel carrier for their use to escape the launchpad area. While they did not ride the slidewire baskets, each astronaut practiced driving the M-113.

The actual TCDT was a dry run of the final phases of countdown—without propellants in the tanks—with the crew aboard the shuttle and my launch team and me in the Firing Room at the Launch Control Center. The TCDT stopped at T minus five seconds in the countdown.

The crew then emerged from the vehicle, confident and ready to fly the mission. They posed on the launch tower's highest access arm for a traditional photo with their shuttle in the background. Robert Hanley was at the pad, monitoring activities during the TCDT. He asked the KSC photographer to take a picture of him with the crew. That photo became one of Hanley's most cherished keepsakes.

Traditions are an important morale builder in a program as long-lived as the shuttle. One TCDT-week tradition was for the Astronaut Office to host a dinner for the flight crew and some invited guests at the astronaut beach house, located on the shore a few miles south of the launchpad. It was an opportunity for the crew and about a dozen NASA and contractor managers from KSC to get to know one another and unwind a bit. Through the managers, the astronauts could pass along their thanks to all of the team members involved in checking out, preparing, and launching the shuttle.

NASA provided the food, which was always the same—barbecued smoked sausage and beef brisket, fried chicken, coleslaw, baked beans, potato salad with hard-boiled eggs, and bread and butter. A bowl of sliced jalapeño peppers was available for people who wanted to spice up their food. Dessert consisted of brownies. The crew personally provided the adult beverages; NASA couldn't purchase those with government funds.

I found myself eating with Ilan Ramon. Seeing that he was mostly just picking at his food, I asked, "Are you all right? It doesn't seem like you're enjoying your meal."

Ramon replied, "No, no, it's very good. It's not kosher . . . but it's very good!"

—

Weather in the Houston area was stormy on Sunday, January 12, 2003, as *Columbia*'s crew prepared to fly from Houston to KSC. Rick Husband decided it would be safer and more comfortable for the crew to ride together in NASA's Gulfstream G2 trainer airplane rather than flying out in four of their two-seat T-38 jets. Astronaut Jerry Ross flew out to KSC with *Columbia*'s crew. I met them at the Shuttle Landing Facility runway with my traditional greeting, "Welcome to Launch Week!"

One of my responsibilities was to give the crew a complete security briefing and review security procedures with them. The crew needed to feel absolutely confident about how we would keep them safe on launch day.

The space shuttle was a high-value and highly symbolic national asset, carefully protected by NASA and the US military. Sixteen months after the terror attacks of September 11, 2001, the United States was at war in Afghanistan, and we had an Israeli astronaut on the mission. STS-107 garnered the most stringent security ever implemented for a space shuttle launch.

Security at Kennedy was primarily aimed at protecting the public from NASA's rockets, rather than the other way around. We established a three-mile "box" in the waters off KSC—an exclusion zone to keep aircraft and boats out of the launch path in case of an explosion early in a rocket's flight. But now we also had to consider the very real possibility that the shuttle could be attacked.

If a plane or boat strayed into the restricted zone around KSC and the vehicle flight path during a countdown, we faced tough decisions. Was it a tourist who just wanted to take some photos up close? Was it a charter fishing boat that strayed off course and forgot to turn on its radio? Or was it someone trying to look innocent, only to then make a sudden hostile move? As the launch director, I had to decide in the moment whether to tell the crew to sit tight in the shuttle or direct them to make an emergency escape. Calling for the escape assured the safety of the crew but could damage the shuttle in the process, forcing a long turnaround before the next launch attempt.

To thwart potential terror attacks, we kept *Columbia*'s scheduled launch time a secret in the weeks leading up to the mission. We had even briefly considered a scheme dubbed *Operation Yankee*, which would have entailed a surprise liftoff one day in advance of a publicly announced launch date.

Finally, on Wednesday, January 15, we announced that "T-zero" for *Columbia*'s launch would be at 10:39 Eastern Time the next morning.

The highly publicized mission drew large crowds to the KSC area. NASA's public affairs office requested that more spectators than usual be allowed on site. KSC security went into round-the-clock operations. VIPs and other spectators parked at the KSC Visitor Complex and boarded buses to the viewing stands. Crowd control and protecting the public were the order of the day. In case of a launch emergency, security would have to get all spectators onto buses as quickly as possible for their own safety.

As launch director, I usually pulled a twelve-hour shift on launch day. Officially, I had to be on duty as the shuttle's external tank was loaded with propellants. That operation began at T minus six hours in the countdown, which was actually about nine hours before launch because of the built-in hold periods in the countdown. But there was always a weather briefing an hour before propellant loading could begin. The launch team needed to consider not only the weather forecast at KSC at the scheduled time of liftoff, but also the weather at potential trans-Atlantic abort landing sites in Europe and Africa. There was no use spending the time and resources to load the shuttle's tanks if it appeared that weather restrictions would be violated at launch time. These weather forecasts were part of a larger meeting meant to ensure that everything was ready for fueling to begin and for the mission to fly. It made for a very long day.

Firing Room 4 of the Launch Control Center was already a hive of activity when I arrived the night before the launch. Roughly 180 engineers and managers controlled the countdown activities from that room. They were supported by about as many people in the backup Firing Room—the systems experts who knew the vehicle and ground support systems better than anyone. They were on hand to help resolve the usual technical glitches that cropped up during the countdown. VIPs and the prelaunch Mission Management Team observed the proceedings from the two glass-walled "bubbles" flanking the top row of consoles and my station.

A multiagency command center was created for this mission and operated from the second floor of the Launch Control Center. FBI, CIA, and state and local security forces staffed the command center and monitored the operations. Mark Borsi, director of KSC security, reported directly to me on the status of security measures and issues. I also had direct access to coast guard and air force brass via my console.

After two years of rescheduling and delays, launch day for STS-107 finally dawned on January 16, 2003.

In the early morning hours, ground systems pumped 146,000 gallons of liquid oxygen and 396,000 gallons of liquid hydrogen into the shuttle's external tank. At T minus three hours in the countdown, we deployed the Ice Team to the launchpad for a two-hour inspection of *Columbia* and the pad systems. The Ice Team used binoculars, telephoto lenses, and infrared devices to check for any unusual ice buildup or other debris on the surface of the vehicle, because the super-cooled liquid propellants in the tank might cause ice to form from the always-humid Florida Coast air. A chunk of ice falling off the tank or a booster and striking the orbiter during the ride into orbit could doom the shuttle. The team saw nothing unusual, and they reported their findings back to us in the Firing Room. The countdown proceeded relatively smoothly.

In the Operations and Checkout Building eight miles from the launchpad, astronauts Jerry Ross, Kent "Rommel" Rominger (chief of the astronaut corps), and Bob Cabana (director of flight crew operations) were on hand as *Columbia*'s crew suited up before the flight. Robert Hanley, who reported to Ross, filmed the proceedings with Dave Brown's video camera. Brown requested that Hanley videotape the suit-up and walkout as part of the commemorative video that Brown was compiling about the crew's training. As the crew and their entourage walked out of the building toward the waiting Astrovan, Rick Husband and Willie McCool reached overhead and touched their hands to the STS-107 mission decal on the head of the door frame. It was yet another of the good-luck traditions for space travelers.

The Astrovan drove north to the corner of Kennedy Parkway and Saturn Causeway, near the Vehicle Assembly Building. There, the van stopped to let out Rominger, who went to the Shuttle Landing Facility. He would then fly the Shuttle Training Aircraft around the KSC area to monitor weather conditions throughout the remainder of the launch countdown. NASA needed firsthand accounts of visibility and winds aloft, in case *Columbia* needed to make an emergency return to the runway. Cabana, Ross, and Hanley said their good-byes to the crew and left the van at the checkpoint by the Launch Control Center. Ross and Hanley joined the crew families in the Launch Control Center. Cabana went to the "bubble" adjacent to the Firing Room to join the Mission Management Team. The flight surgeon came in to man the Firing Room's biomedical console.

The STS-107 crew rode the remaining three and a half miles to the pad. After pausing for a quick look up at *Columbia*, they took the short elevator ride to the 195-foot level on the Fixed Service Structure, and walked across the crew access arm to the White Room. Over the next fifty minutes, the pad closeout crew strapped the astronauts into their seats, and then sealed the hatch. The astronauts went through their checklists for the final stages of the launch countdown.

—

The children of *Columbia*'s crew busied themselves drawing with markers on a whiteboard near my office on the fourth floor of the LCC as the final hours of the countdown ticked by. It was another of the KSC launch traditions—a way to keep the kids occupied during what would otherwise be a tedious time for them and to afford their parents some time to be alone with their thoughts. In the days following the launch, our staff would frame the children's whiteboard art and mount it in the hallway to join the scores of "kid pics" from previous missions.

When the countdown came out of the final scheduled hold at T minus nine minutes, escorts took the children and the rest of the immediate members of the crew's families out onto the LCC roof and up a stairway to a private viewing area. There they could watch the launch, shielded from the eyes of the press and public—a precaution we implemented after the *Challenger* disaster.

In the Firing Room two floors below, my launch team prepared to come out of the hold. It was a final chance for managers and engineers to catch their breath and work any last-minute issues. Things would move very rapidly once the count started up again, almost entirely under control of the ground launch sequencing computers.

We were not nervous, but the atmosphere was charged and intense. The room was dead quiet. Launching a space shuttle is *never* routine.

That intensity went to a whole new level for me when I received a call on the secure line from the air force. Their tracking radar showed an unidentified object due south of the launch complex, heading due north.

Estimated time of arrival at the launchpad area: T-0.

*Holy shit. This is it. We're under attack.*

I received several reports of the object's position, and the command center team plotted its course. Then it disappeared from the radar screen. It reappeared and disappeared several more times over the next couple of minutes.

*What the hell is going on?*

I called the general who was my liaison at the Department of Defense. He had a direct line to the president in case authorization was needed to shoot down a civilian airplane. The general relayed to me what the pilot of the air force jet circling overhead was saying. Secondhand information always has the possibility of being misinterpreted. I said, "Sir, I trust you completely, but I need to speak to the pilot directly." The general objected at first, as this was a breach of protocol between the military and a civilian agency. I insisted, and a few seconds later, they patched me through to the pilot.

My hands were shaking. I held onto my console to steady myself. I asked the pilot, "Sir, if there was anything out there, would you see it?"

The pilot responded, "Yes, sir, I would."

"And do you see anything?"

"No, sir, I don't."

I was listening carefully to the pilot's choice of words and the tone of his voice. If he had said, "I don't *think* I see anything," or if there had been any hint of uncertainty in his voice, I probably would have told the crew to punch out, ordering them to make an emergency egress. However, the pilot sounded completely confident.

All the months of planning, the security exercises, the resources deployed, the years of the crew waiting for the mission to fly, the relentless training, the scheduling pressure to fly this mission—it all came down to this decision.

*Was there an emergency or not?*

At the end of the final countdown hold, my tradition was always to give the crew an upbeat send-off message on behalf of the launch team. I got on the comm loop with Rick Husband and said, "If there ever was a time to use the phrase, 'Good things come to people who wait,' this is the one time. From the many, many people who put this mission together: Good luck and Godspeed."

Rick replied, "We appreciate it, Mike. The Lord has blessed us with a beautiful day here, and we're going to have a great mission. We're ready to go."

I gave the "Go" for the count to pick up on schedule.

The fighter pilot had assured me there was no visible threat, but the internal voice of doubt nagged me.

*God—what if I'm wrong?*

The final minutes of the countdown quickly ticked away, and all went smoothly. I nervously looked out the window toward the launchpad every few seconds, half expecting to see something heading toward *Columbia*.

At the launchpad, everything was proceeding exactly as planned. A few seconds before 10:39 a.m., *Columbia*'s three main engines ignited and quickly built up to steady thrust. *Columbia*'s nose rocked forward several feet in reaction to the off-center impulse from the engines and buildup of thrust two hundred feet below. The instant the shuttle rocked back to vertical again—6.6 seconds after main engine ignition—the twin solid rocket engines fired. Explosives shattered the hold-down bolts at the same moment, and *Columbia* leaped into the clear blue sky. Launch and Entry Flight Director LeRoy Cain at Mission Control in Houston assumed control of the mission as soon as the solid rocket boosters fired.

I breathed a deep sigh of relief.

—

The families filed back into the LCC after *Columbia* disappeared from sight about two minutes after liftoff. Eight and a half minutes after launch, *Columbia* was in orbit. It seemed to be a picture-perfect launch.

After the postlaunch checklists were complete, the launch team and I went to the lobby of the Launch Control Center for the celebratory meal of beans and corn bread, which was served after every successful launch. Still too on-edge to eat, I took a quick bite and shook hands with a few folks on my way to another tradition—the postlaunch press conference. Fortunately, no one else present there knew about the security incident. And they didn't need to know.

It was later determined that the unidentified object on radar was a cluster of Mylar party balloons with a small, empty metal box—about the size of a clock radio—dangling underneath. Riding on the winds, the balloons dipped into and out of radar coverage. They were found two days later on the shore of the Banana River approximately five miles south of the launchpad.

## Chapter 3

# THE FOAM STRIKE

Once *Columbia*'s main engines shut down, the flight computer commanded pyrotechnic charges to fire to jettison the external fuel tank. Astronaut Mike Anderson triggered cameras on the shuttle's belly to take photos of the tank as the shuttle pulsed its maneuvering thrusters to move away. Those photos were part of the launch documentation, to note any issues that might require attention on the next missions. The crew did not notice anything unusual about the tank as it slowly drifted away from them. As usual, the tank would break up and fall into the Pacific Ocean south of Hawaii.

Standard procedure called for the tank photos to be transmitted to the ground at the end of the first day's operations. However, the *Columbia* crew had a busy day ahead of them configuring the experiments aboard Spacehab.

The photos of the tank were never downlinked.

If engineers on the ground had seen the photos, they would have immediately noticed that a large piece of foam—about the size of a carry-on suitcase—was missing from the area at the base of the left side of the strut connecting the orbiter's nose to the tank.

Back on the ground, an array of cameras along Florida's Space Coast had filmed *Columbia* on her ride uphill. The imagery analysis team at KSC began reviewing the films the afternoon after the launch. The team was frustrated to discover that one of the tracking cameras had not worked at all, and another was out of focus.

What particularly caught their eye, however, was footage from one camera showing what appeared to be a large piece of foam falling off the tank 81.7 seconds into the flight. It fell toward the *Columbia*'s left wing and then disintegrated into a shower of particles. The foam had clearly struck the orbiter, but it was impossible to tell from the images exactly where it had impacted or how bad any damage might be.

Ann Micklos, who represented her thermal protection system team during the video review recalled that "people's jaws dropped. You could have heard a pin drop in that room when we saw the foam strike. We watched it on the big screen again and again and again, trying to understand where the foam impacted the orbiter."

It was indeed an impressive-looking impact, but debates about its severity began almost immediately. Was this a serious situation? Or was it like all the other impacts—posing a maintenance inconvenience but not a threat to the crew? The imagery lab in Tower K of the Vehicle Assembly Building went to work to enhance the video as much as possible.

Why is foam shedding even a concern? To understand that, we need to review the vulnerability of the shuttle's design. The shuttle's flexibility was ironically its biggest downfall. Unlike previous spacecraft designs for Mercury, Gemini, and Apollo—in which the capsule with the astronauts was at the front end of the rocket—the space shuttle and its crew rode into orbit *beside* the propellant tank and the rocket boosters. This meant that ice or other debris could fall off the tank and boosters and strike the shuttle during ascent. Damage from launch debris was one of NASA's major headaches, as there was no way to repair an orbiter's exterior surfaces once the vehicle reached orbit.

We did not want to put a wounded space shuttle into orbit if we could avoid doing so.

Chief among the concerns was the intricate heatshield system completely covering the orbiter. The shuttles had aluminum skin, and when "naked," they looked remarkably similar to conventional aircraft. However, aluminum has a relatively low melting point and cannot withstand the blazing temperatures of reentry. NASA's ingenious heatshield for the shuttle consisted mostly of a system of silica tiles, which not only insulated the vehicle's structure, but actually radiated heat away from the shuttle. The tiles were lightweight, porous, and crumbled easily. They covered the belly, the tail, and the maneuvering engine pods protruding from the aft end of the vehicle. The tiles could not be applied as a single unit or even a few large pieces, because the orbiter's airframe had to flex during launch and reentry as it encountered air resistance. So, the tile system ended up being a mosaic of thousands of tiles, each approximately six inches square and each with a unique shape. Each relatively fragile tile was glued to a felt pad, which was itself glued directly onto the aluminum skin of the orbiter. This allowed for slight movement in the orbiter's structure without damaging the tiles. Every tile was numbered so that it could be readily identified and placed in the appropriate spot on the orbiter.

The tiles were not the only components of the orbiter's heatshield. Some parts of the shuttle were exposed to more extreme heat than the tiles alone could withstand. The nose cap of the orbiter and the leading edge of its wings were made of a dark gray reinforced carbon-carbon (RCC) material that could withstand heat of up to 3,000°F. RCC was hard but brittle. Other parts of the shuttle, which were subject to much less heat during reentry, were covered with quilt-like blankets of silica and felt.

If the foam impact we saw had severely damaged the tiles on *Columbia*'s belly or impacted the wing's leading edge, searing hot plasma could enter the vehicle during reentry and melt the ship's internal structure.

There was no backup to the heatshield system. If it was seriously compromised, the crew was not going to make it home.

—

NASA created the Mission Management Team (MMT) process after the *Challenger* disaster as a way to ensure that potential issues like

*Challenger*'s O-rings[1] came to the attention of Shuttle Program managers. NASA wanted a way for information to flow quickly to senior management, without being filtered or suppressed. The "prelaunch" MMT sat on the same row as me in one of the glass-walled "bubbles" in the Firing Room on launch day listening for anything that might make them question launching that day. It was their job to assess issues that were not part of the documented launch countdown process but that could still pose risks to the crew and thus be reasons not to launch.

Many of the members of the prelaunch MMT then moved on to the "on-orbit" MMT. This MMT was supposed to meet regularly during a mission to assess any issues that arose during flight or could affect landing.

Kennedy Space Center's operational responsibility for a mission essentially ended once the shuttle had blasted off. However, KSC managers—including me, as launch director—participated in the on-orbit MMT. We focused on any issues that might affect preparations for the vehicle's next mission once the shuttle returned from flight. Had something cropped up that would delay *Columbia*'s processing flow for her next flight? Were there any special issues that would require unusual servicing on the runway?

The imagery analysis team reported the foam strike to the MMT on January 17, the second day of *Columbia*'s mission. Their initial conclusions were unprecedented. No one had ever seen such a large piece of foam come off the external tank and impact the orbiter during ascent. The analysts said it was a "big hit," but no one knew how bad it was. It was just lightweight foam, but it may have hit the orbiter in a potentially dangerous location. The MMT did not appear concerned, but they asked for further analysis.

Engineering teams examined the limited data they had at their disposal. The only computer program available was designed to model the impact of ice particles on the tiles on the orbiter's underside. The software algorithms were not intended to assess damage from foam insulation strikes or hits on areas other than tiles. With so many unknowns, it was difficult to get consistent results from the analyses. The software's appropriateness for use in this case was a stretch, but it was all that our analysts had to work with.

The consequences of foam impacts on the RCC material on the leading edge of the wing were even more of an unknown. Reinforced carbon-carbon had a very hard surface, which some specialists considered too tough to be seriously damaged by foam impacts. Some engineers tried unsuccessfully several times to convince their colleagues that RCC was less forgiving than tile, but their objections largely went unappreciated. While someone could almost crush tiles in their hands, RCC felt like an extremely tough and capable material.

Despite the application of the best minds to the problem, there was simply no reliable way to predict what the damage might be. Ann Micklos said, "We never had a clear picture of where the impact was. It was all assumptions. And you can't solve a problem based on assumptions."

Robert Hanley and his boss Jerry Ross had returned to Houston after the launch. Hanley mentioned to Ross in a hallway conversation that he had heard reports about people investigating a possible debris strike on *Columbia*. Hanley said the word going around JSC was that the foam strike was a "nonissue." Everyone thought foam was too lightweight to cause any serious damage.

Ross replied to Hanley, "I'm not so sure."

Ross was recalling his experience as a crewman on *Atlantis*'s STS-27 mission, where hundreds of the orbiter's tiles were heavily damaged during launch, and missing tile created a hole in the heatshield that nearly burned through on reentry. *Atlantis* held the distinction of being the most heavily damaged spaceship ever to survive reentry.[2] And on STS-112, just four months before *Columbia*'s launch, a smaller piece of foam fell off the external tank and dented the metal ring attaching the left solid rocket booster (SRB) to the tank.

The MMT discussed the foam strike at the four MMT meetings held during *Columbia*'s mission. The engineer presenting the issue to the MMT was new to his position. The MMT pressed him on data to back up his conclusions about potential damage to *Columbia*—in essence, "prove to us there's a problem." He responded that the team needed more data to make an accurate assessment.

Engineers who had been with NASA since the early days of the shuttle recalled that national security assets had been called into service to photograph *Columbia* in orbit on her maiden flight in 1981. On that mission, as soon as the payload bay doors were opened once *Columbia* reached orbit, the crew and NASA noticed several tiles missing from the area near the shuttle's tail. It was not public knowledge at the time, but NASA had reached out to the intelligence community to take images of *Columbia* in orbit and determine if other tiles were missing in critical areas not visible to the crew. How those images were obtained, and what they were able to show, is still classified information.

People who remembered that situation reached out to contacts in the intelligence community and asked if it was possible to take similar images now. Mid-level technical experts in the intelligence community said they would be happy to help. They just needed a formal request from NASA.

In one of the most confounding breakdowns of the management process for STS-107, the MMT refused to issue a formal request for images. In essence, the reasoning was: "You don't have enough data on the problem to warrant getting the intelligence community involved." And yet there was no way for the team to gather more data without the intelligence imagery. The imaging capabilities possessed by the intelligence community were highly classified and could not be used as justification for the request, because most of the team was not cleared to hear that information. Trapped in a Catch-22, those who desperately wanted the additional information felt incredibly frustrated at the bureaucratic logjam.

Lower-level engineers at KSC and at Boeing's shuttle design offices in Huntington Beach, California, adamantly insisted that the foam impact had damaged the RCC. Some refused to certify that the vehicle was safe to come home. The MMT noted and then overruled their objections. The discussions were not even reflected in the MMT's meeting minutes.

The MMT members convinced themselves there was nothing to be overly concerned about for *Columbia*'s reentry. Rather than digging into the possibilities of what could go wrong, they reassured one another that everything would be all right.

"Prove to me that it's *not* safe to come home" demonstrates a very different management culture than does "prove to me that it *is* safe to come home." The former attitude quashes arguments and debates when there is no hard evidence to support a concern. It allows people to talk themselves into a false sense of security. The latter encourages exploration of an issue and development of contingencies.

In hindsight, many of us who participated in the debate and decisions—myself included—blamed ourselves for not pressing the issue about the foam strike. However, it simply did not occur to most of us at the time that the crew might be in danger. Complacency and past experience lulled us into believing that the shuttle would get her crew home safely—just as she had done more than one hundred times previously—despite the knocks and dings. Management assumed that if there really were a problem, the "smart people" who were looking at it would speak up. Managers seemed not to comprehend that objections *had* in fact been raised and then brushed aside. Pressing on with the mission so that NASA could get back to space station assembly flights just seemed like the right thing to do.

The crew was not even told about the foam strike until January 23—one week into the mission—and then only to prepare them for a question that might arise in an upcoming press conference. Mission Control sent an email to Rick Husband and Willie McCool informing them about the hit and immediately downplaying any worries: "Experts have reviewed the high speed photography and there is no concern for RCC or tile damage. We have seen this same phenomenon on several other flights and there is absolutely no concern for entry."[3]

Even if the astronauts had been asked to look for damage, they could not have shed light on the situation without taking extraordinary measures. Most of the front of the wing was not visible from the windows in the cockpit. The orbiter was not carrying its robotic arm in the payload bay, because the arm was not needed for this mission. Had that arm, with its multiple television cameras, been available, the crew could have scanned the top and front of the wing for damage. Even so, the arm would not have been able to reach underneath the orbiter to look for damage there.

The only other way the crew could have checked the wing for damage would have been to take a space walk. That would have required a two-day interruption to the science activities in Spacehab. The pressurized tunnel to Spacehab was on the other side of the air lock in the crew compartment. The crew would have had to seal off Spacehab while preparing for and conducting the space walk.

So, the MMT did not ask the crew to inspect the orbiter. The MMT incorrectly concluded that no significant damage existed. Besides, the MMT reasoned, there was nothing the crew could have done about it anyway. The MMT flatly declared that there was no "safety of flight" issue involved—that is, no risk for reentry. Any damage to the thermal protection system would just be a turnaround maintenance problem for the next mission once *Columbia* was back on the ground.

The US Air Force's Maui Optical and Supercomputer Site (AMOS) took images of *Columbia* as it passed over Hawaii on January 28. The orbiter's payload bay was facing the cameras on the ground. The Spacehab module was clearly visible in the payload bay. Unfortunately, the open bay doors obstructed the view of the front half of *Columbia*'s wing, where the foam was thought to have struck the ship. The resolution of the AMOS cameras was probably not good enough to have captured wing damage anyway.

That same day—the seventeenth anniversary of the *Challenger* accident—Rick Husband and his crew paused to remember the crews of *Challenger* and Apollo 1. Husband said, "They made the ultimate sacrifice, giving their lives in service to their country and for all mankind. Their dedication and devotion to the exploration of space was an inspiration to each of us and still motivates people around the world to achieve great things in service to others. As we orbit the Earth, we will join the entire NASA family for a moment of silence in their memory. Our thoughts and prayers go to their families as well."[4]

—

Unknown to NASA at the time—and even to the people manning the intelligence assets that acquired the images—the US military had inadvertently obtained evidence of something breaking away from *Columbia* on the second day of her flight.

The Space Surveillance Network (SSN), which was operated jointly by the US Army, Navy, and Air Force, was a worldwide network of sensing systems designed to track objects in orbit around Earth. Early in the postmortem of the *Columbia* accident, SSN analysts went back over their tracking data to see if they had obtained any information about *Columbia* and any objects that might have collided with her in orbit. The analysts noticed that another object was in the same orbit as *Columbia* beginning on the second day of the mission.

After refining the radar data, the analysts determined that a slow-moving object, about the size of a laptop computer, gradually drifted away from the shuttle. Its slow motion implied that it was probably not a piece of space junk or a meteor. Further tests showed that the radar properties of the object were a close match for a piece of RCC panel—possibly part of the wing's leading edge. It appeared to separate from the shuttle after several thruster firings that changed *Columbia*'s orbital orientation.

Whatever it was, the object reentered the Earth's atmosphere and burned up on January 20, twelve days before the end of *Columbia*'s mission. Theories about the object and its origin were debated at length during the accident investigation, but its exact nature and possible relevance to *Columbia*'s demise will never be known.

Again, no one knew anything about this object during the mission. Could this information have changed the course of events? That will also never be known.

—

As *Columbia* approached the end of her time in orbit, some people at KSC began to worry about how to bring her home safely. If the thermal tiles had significant damage, *Columbia* would need to keep its temperature down as much as possible during reentry.

Weight was an immediate concern. As the first ship in the fleet, *Columbia* was already heavier than her sister shuttles. The added mass of the Spacehab module meant that STS-107 would be the heaviest shuttle ever to return from orbit. That would make her reentry hotter than usual, even if everything went as planned.

Some people at KSC openly asked, "Can we jettison the payload to make the vehicle lighter?" Even if that were possible—and it was not—it would have meant the loss of many of the science experiments and their data. Management did not seriously consider the recommendations to throw overboard all "loose objects" in the crew module and Spacehab, especially with the official determination that there was no concern for flight safety.[5]

Ed Mango, my assistant launch director, had supported "Hoot" Gibson and Jerry Ross's STS-27 mission early in his career at NASA. Recalling how badly beat up *Atlantis* was on that mission, Mango expected that *Columbia* would also make it back to the landing site, but would probably be heavily damaged. He requested permission in advance to go out to the runway after the vehicle was "safed" so that he could inspect the ship personally.

Ann Micklos said, "Putting on my engineering hat, we were interested to see what the vehicle was going to look like when it came back. I thought she could handle reentry with the damage. I trusted the team that provided the 'go for entry.' However, our team was also planning how we were going to turn the vehicle around on the ground to get it ready for [missions to the International Space] Station. None of us in our wildest thoughts believed that things would turn out the way they did."

Pressure to keep on schedule had combined with a complacency brought about by so many past mission successes. The same conditions were present for Apollo 1 and *Challenger*. And once again, a crew would pay with their lives.

## Chapter 4

# LANDING DAY

### 8:15 a.m. EST

As Saturday, February 1, 2003, dawned, NASA teams at Kennedy Space Center and Johnson Space Center prepared to bring *Columbia* home at the end of her sixteen-day mission. Orbiting 173 miles above Earth,[1] *Columbia*'s crew stowed their experiments in lockers and closed and latched *Columbia*'s payload bay doors. The crew donned their launch and entry pressure suits and helmets. These orange "pumpkin" suits could provide protection and oxygen in case the cabin lost pressure during reentry or the crew had to make an emergency bailout.[2]

The astronauts took their positions in the crew module. On the flight deck, Rick Husband sat in the commander's seat, with pilot Willie McCool to his right. Kalpana Chawla sat in the jump seat, behind and centered between Husband and McCool. To her right was Laurel Clark. These four astronauts had the best view during reentry, as all of the shuttle's large windows were on the flight deck. The three crewmen seated in the shuttle's mid-deck (seated from left, Mike Anderson, Dave Brown, and Ilan Ramon) could see only the stowage lockers directly in front of them.

*Columbia* could land either at Kennedy or at Edwards Air Force Base in California, depending on the weather in Florida. NASA had to make the landing site decision about two hours in advance, because the de-orbit burn needed to happen halfway around the world from the landing site and one hour before touchdown.

The weather forecast at Kennedy looked favorable, so Commander Husband was given the "Go" to come home to KSC. He fired *Columbia*'s orbital maneuvering system engines over the Pacific Ocean at 8:15:30. This would drop the ship out of orbit and direct her toward Florida, with a touchdown at 9:16 Eastern Time. She would come in to Kennedy's Shuttle Landing Facility—the SLF—one of the world's longest runways.

*Columbia* gradually fell out of orbit and became an unpowered glider in the atmosphere. Now irrevocably committed to land at KSC, there was no option to go around again or look for another spaceport.

Launch and Entry Flight Director LeRoy Cain and his Flight Control team in Houston were managing the last stages of the flight. They monitored telemetry from the vehicle and communicated with the crew. They would continue to command the mission until *Columbia*'s crew disembarked at Kennedy.

Outside Mission Control, the rest of Johnson Space Center was almost a ghost town as this cool, foggy day dawned. People who were not involved in mission operations for *Columbia* or the International Space Station were at home for the weekend. Andy Thomas, deputy chief of the astronaut office, was on routine duty in Building 4 South that morning, manning the contingency action center. There would normally be very little for him to do.

I arrived at Kennedy's Launch Control Center at seven o'clock to monitor the de-orbit burn. Ed Mango was there, as was Don Hamel, controlling Kennedy's landing operations. Mango came in that morning to fill in for a vacationing colleague. He told his wife to expect him home by ten-thirty, after *Columbia* was securely in the hands of the landing and recovery team. Our main role as managers was to deploy the landing and recovery forces to assist the crew and power down *Columbia* after she landed. Nearly two hundred people crowded into the Firing Rooms on launch day. But on landing day, things felt almost deserted, with only about fifty people on hand.

I watched the proceedings on one of my computer monitors, and I saw that the de-orbit burn occurred on schedule and without problems. I then raced out to the Shuttle Landing Facility runway, about two miles north, to join other VIPs and guests gathering to greet *Columbia.* I did not want to get stuck waiting for the convoy of landing support vehicles to trundle across the roadway. They had also departed their staging area for the runway when they heard that the de-orbit burn was successful.

*Columbia* began encountering the upper reaches of the atmosphere. Her path would take her east-southeast across California, Nevada, Utah, Arizona, New Mexico, Texas, and Louisiana. Then she would skirt the Gulf Coast as she lined up for the approach to the runway.

As she made her way across the southern United States over the course of twenty minutes, *Columbia* would lose her last forty-four miles in altitude and slow from sixteen thousand miles per hour to zero at "wheels stop." *Columbia*'s automated flight control system ran all aspects of the flight during reentry, banking and rolling the orbiter to control its speed and energy profile during the period of maximum heating. Rick Husband would take the control stick only in the final couple of minutes of the approach and landing sequence.

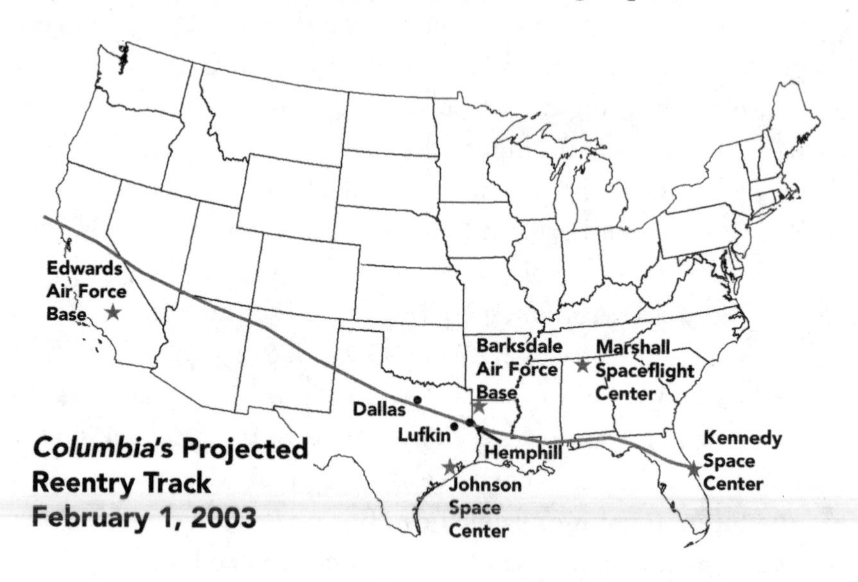


*Columbia*'s projected reentry path over the United States.
Contact was lost with the ship just after it passed south of Dallas.

## 8:30 a.m. EST

Landing day at Kennedy Space Center was invariably a happy occasion, and the mood was much less tense than for launch. While both launches and landings were cause for celebration, crew families understandably held their breath and prayed during a shuttle's ascent to orbit. It was impossible to forget the *Challenger* accident, so launches were always accompanied by unspoken fears that the families might never see their loved ones again. In contrast, landing day meant eager anticipation for the approach of the shuttle, followed by jubilation, pride, and relief when the orbiter's wheels came to a stop on the runway.

Other than one blown tire and one touchdown just short of the runway—fortunately, at the dry lake bed at Edwards—there had never been a problem with the previous 111 shuttle landings. Rick Husband and Willie McCool—two of the world's best pilots—had practiced the landing approach more than one thousand times in simulators and the Shuttle Training Aircraft. No one worried about their making anything less than a perfect landing.

The entourage of VIPs, crew families, and other support personnel began arriving at the SLF's midfield park site to await *Columbia*'s arrival. NASA Administrator Sean O'Keefe sported a red STS-107 polo shirt. Former shuttle astronaut Bill Readdy, who was O'Keefe's associate administrator, accompanied him at the runway, as did KSC Center Director Roy Bridges (also a former astronaut) and Paul Pastorek, NASA's chief counsel.

Our landing and recovery team had been working since five o'clock to prepare the service vehicles that would meet *Columbia*. The "go for de-orbit burn" call was their signal to deploy the convoy to the runway. Once the shuttle landed, the team would "safe" the orbiter by checking for the presence of hazardous propellant fumes. Then they would power down the systems and help the wobblier astronauts into the crew transport vehicle, which was similar to the mobile lounges at some airports.

My launch director role at the runway was largely ceremonial. I would have the honor of welcoming the crew home after they exited the vehicle. The crew members typically spent twenty minutes walking around to inspect the orbiter—its tiles still radiating warmth from reentry—and to thank the KSC workers. The astronauts would say a few words to the press and then board the Astrovan to meet up with their families at the crew quarters. The VIPs and I would congratulate one another on the conclusion of a successful mission. Then we would conduct the traditional postlanding news conference at Kennedy's press site.

The recovery convoy was deployed as usual, with half the vehicles at one end of the SLF and half at the other end. The shuttle could alter its approach direction any time during the final ten minutes depending on the wind direction, so teams waited at both ends of the runway until the orbiter landed and came to a full stop. This morning's light breeze from the west-northwest meant that *Columbia* would most likely make her final approach from the south-southeast.

Astronaut Jerry Ross stood by the convoy command vehicle. With him was astronaut Pam Melroy, the pilot of October 2002's STS-112 mission. Just back from a trip to England, Melroy was about to take on the role of "Cape Crusader," an astronaut supporting the crew of the next shuttle mission at KSC. She was at the runway as part of her familiarization training, to remind her of the steps involved in unstrapping the crew and taking over the cockpit from them.

At 54°F, the morning was cool for Florida but not uncomfortable. I scanned the sky and asked the KSC weather officer, John Madura, if the slowly building clouds were a concern. "They'll be all right," Madura said. "They'll come through some clouds, but they'll see the runway."

**8:54 a.m. EST**

During a shuttle mission, NASA always kept a landing and recovery team on standby at Edwards in case the orbiter needed to land there. Robert Hanley was in California with the standby team. Since the shuttle was now headed toward KSC, Hanley was off the hook for the rest of the day. He could watch the proceedings on TV at NASA's Dryden Flight Research Center, which was on the grounds at Edwards. He would clean up some paperwork and then catch a commercial flight home to Houston.

Hanley knew that the shuttle's reentry path would take it over Edwards. The ship's blazing plasma trail would be a spectacular sight in the predawn sky over the high desert. Hanley and a companion pulled their car off the road en route to Dryden. He phoned his mother, who was monitoring *Columbia*'s reentry on TV. She gave him updates as the ship's track approached the California coast. At 8:54 EST, she told him, "It's coming up! It's coming up!"

Streaking across the predawn sky at 15,500 mph and at an altitude of 230,000 feet, *Columbia* was a fast-moving, breathtakingly bright "star" followed by a beautiful glowing pink and magenta trail of ionized oxygen. Transiting the sky in only a minute, the shuttle blazed off to the southeast over Nevada and Utah.

Hanley got back in his car and raced on toward Dryden. He had twenty minutes to get to his work trailer so that he could watch the NASA-TV feed of *Columbia*'s landing in Florida.

He did not know that he was among the last of the NASA family to see *Columbia* in flight.

### 9:00 a.m. EST

At almost precisely the same time Hanley watched *Columbia* fly over California, flight controllers in Houston began receiving unusual telemetry readings from the orbiter. Temperature readings from four sensors in *Columbia*'s left wing began to rise. Then the sensors went dead within a few seconds of one another at 8:53. At 8:58, as *Columbia* crossed the New Mexico-Texas line, the tire pressure readings in *Columbia*'s left landing gear started to look unusual. Then those sensors also dropped off-line.

Ed Mango was monitoring the flight controllers' conversations from Kennedy's Firing Room. He thought it odd that these unrelated sensors would all start failing at about the same time. He became uneasy. Something was not right. The sensors implied unexpected heat inside the wing. However, the status displays showed that the shuttle appeared to be flying its programmed S-turns normally.

*Columbia* was above Dallas at 8:59:32 when Commander Rick Husband's communication to Mission Control was cut off mid-word. Mission Control also stopped receiving telemetry from *Columbia* at that instant.

Occasional communications dropouts were not unusual during reentry, because the ionized plasma sheath that was building around the shuttle sometimes disrupted radio signals. However, this blackout lasted much longer than expected. After a few minutes, Mission Control's astronaut communicator, Charles Hobaugh, attempted to raise *Columbia* several times. His repeated calls of "*Columbia*, Houston, comm check" went unanswered. Long periods of silence ticked by between his calls.

Mango knew something was seriously wrong when Hobaugh switched to the backup UHF radio system to try to raise the crew. The tracking radar in Florida was also not picking up *Columbia*. The ship should have appeared over Kennedy's radar horizon by now.

The first thought that crossed Mango's mind was: *Ballistic entry. Maybe it's going to try to land at an airport in Louisiana.*

**9:05 a.m. EST**

While we stood beside the runway at Kennedy, residents of East Texas were waking up to a chilly February morning. The sun had not yet burned off the fog enshrouding the dense pine forests of the hilly countryside. Temperatures hovered just above freezing.

Most citizens of that part of the state were unaware that *Columbia* would be passing overhead on its way to Florida that morning. Many did not even know that NASA had a space shuttle in flight. It simply wasn't something that concerned them.

That suddenly changed, just after eight o'clock local time.

FBI special agent Terry Lane lay half-asleep in bed at his home in Douglas, thirteen miles west of Nacogdoches. He thought he was dreaming about an unusual noise. He quickly realized that he was awake. The noise was real. A rumbling sound grew constantly louder and continued for several minutes. By the time he got out of bed and opened his front door, the noise had subsided.

Farther east, in Sabine County, near the Louisiana border, timber sale forester Greg Cohrs of the US Forest Service was also startled from his sleep. He heard a tremendous boom followed by a rumble that lasted for minutes. His wife Sandra put on her housecoat and opened the back door. She heard popping and crackling noises in the air above. Cohrs tried to imagine what could produce such a constant rumbling and banging. It was not a naturally occurring sound—certainly not thunder. His mind turned to worries about terrorism. *Had Houston or New Orleans been destroyed by a nuclear explosion?*

"Brother Fred" Raney, minister at First Baptist Church and captain of the volunteer fire department in the small town of Hemphill, heard such an intense blast that he thought the cross-county gas pipeline passing through Sabine County had ruptured. Hemphill's funeral directors—John "Squeaky" Starr and his son Byron—also believed they heard a pipeline explosion. The constant rattling and booming had a rhythmic quality that sounded almost mechanical.

Elementary school teacher Sunny Whittington was in the barn at Hemphill's youth arena. Her children both had animals entered in the county livestock show, and it was time for the first weigh-in. The open-sided structure began shuddering violently, punctuated by a tremendous noise that sounded to her like "a sonic boom multiplied by a thousand times." People ran out of the arena. Whittington saw dozens of smoke trails, some spiraling and some going straight across the sky. She asked her husband, "Tommy, what's happening?" He speculated that perhaps a plane was crashing or that two planes had collided.

House windows vibrated so intensely that people feared the glass would shatter. Knickknacks fell from shelves and dressers. The nonstop booms lasted several minutes, shaking US Forest Service law officer Doug Hamilton's brick house to its foundations. Absolutely convinced that it was Judgment Day, he opened his front door and prepared to meet Jesus.

In addition to the booms, some residents heard sounds like helicopter blades, as large pieces of metal spun through the air and crashed into the ground. Fishermen on foggy Toledo Bend Reservoir heard things splashing into the water all around them. One large object—estimated by some to be the size of a small car—hit the water at tremendous speed, creating a wave that nearly swamped several boats.

Sabine County sheriff Tom Maddox was at his Hemphill office returning phone calls after being out of town the previous week. It was his son's birthday, and he planned to spend the day with his family. He finished his final call and phoned his wife to say he was on his way home. Suddenly, the building shuddered so violently that he thought the jail's roof had collapsed. As the noise subsided, all five of his phone lines lit up. One citizen reported that a plane had crashed in the north end of the county. Another reported a plane crash in the southern end of the county. The next caller said that the gas pipeline running through the county had exploded. A fourth caller said there was a train derailment in the western part of the county, between Pineland and Bronson. Maddox couldn't believe that these disasters were occurring simultaneously all over the county. *What was going on?*

Hemphill's Pat Smith had just settled down with a cup of coffee and turned on her TV. She saw on the news that *Columbia* would be passing overhead on its way to Florida. She said to her dog, "We might see that!" As she sipped her coffee, she heard an explosion followed by constant rattling. She ran outside. Her dog was running around in circles, barking up at the sky. She saw smoke trails going in every direction. She went back inside after a few minutes and heard on the news that NASA had lost contact with *Columbia*. She felt a lump in her throat when she realized what she had just witnessed.

*Columbia* had come apart in a "catastrophic event" 181,000 feet above Corsicana and Palestine, southeast of Dallas, traveling more than 11,000 mph. As the vehicle broke up, lighter pieces decelerated quickly and floated to earth. Denser objects like the shuttle's main engines continued along a ballistic path at supersonic speed until they impacted the ground farther east.

Each one of the tens of thousands of pieces of debris produced its own sonic boom as it passed overhead.

Wreckage of the broken shuttle—and the remains of her crew—rained down over Texas and Louisiana for the next half hour along a path that was two hundred fifty miles long.[3]

### 9:16 a.m. EST

Jerry Ross stood next to the crew transfer vehicle on the Kennedy Space Center runway. The flight doctors, nurses, suit technicians, astronauts, security, orbiter technicians, and Flight Crew Directorate managers on his team were responsible for helping the crew out of *Columbia* and removing some of the critical equipment from the vehicle soon after it had come to a stop. KSC security specialist Linda Rhode stood next to Ross on the runway. She traditionally challenged herself to try to spot an approaching shuttle in the distance before Ross saw it. The shuttle should just be becoming visible by now, still more than seventy thousand feet high and flying supersonically. Rhode and Ross scanned the skies and waited to hear the shuttle's characteristic double sonic booms. These announced the shuttle's arrival in the area, preceding the landing by about three minutes.

There were no booms at KSC this morning. People searched the skies for *Columbia.*

*Columbia* should have been lining up to land on Runway 33 at 9:12. At that moment in Houston's Mission Control, Mission Operations representative Phil Engelauf received a cell phone call from someone who had seen video on TV of *Columbia*'s plasma trail breaking into multiple streaks in the sky above Dallas. The breakup had apparently happened less than a minute after NASA lost communications with *Columbia* at about nine o'clock.

Engelauf and astronaut Ellen Ochoa walked over to Flight Director LeRoy Cain and spoke to him quietly. Cain collected his emotions. He said a silent prayer, took a deep breath, and instructed the ground control officer in Mission Control: "Lock the doors." He commanded the flight controllers to preserve all their notes and the data on their computers. They were told not to make any outgoing calls.

At Kennedy's runway, someone signaled Ross to step into the convoy command vehicle, where Bob Cabana had just received a call from Houston. After he heard the news, Ross stepped out of the van and said a prayer. He called the astronaut escorts for the crew's families at the midfield viewing stands and told them, "We've most likely lost the vehicle and the crew." He told the escorts to get the families onto their bus and away from the press as quickly as possible and take them to the crew quarters.

Ross called his associates Lauren Lunde and Judy Hooper at the crew quarters and instructed them to get the facility ready for the families immediately. Ross then gathered the rest of his team in the waiting Astrovan. They drove the eight miles south to the crew quarters as fast as the vehicle could go.

People standing at the runway could scarcely process their thoughts. They knew something was dreadfully wrong, but no one had any idea what had happened. The audio feed from Mission Control was the only source of live information, and it was silent.

The landing clock counted down to zero and then began counting up.

KSC director Roy Bridges suddenly felt his stomach drop, "like the Earth had just opened into a big void, and now you're falling into it."

Administrator O'Keefe appeared to be in shock. A roller coaster of emotions swept over him. He swung from elation at the prospect of greeting the crew to the very depths of despair as he looked at the crew families and realized the horror they were experiencing. He knew *Columbia*'s loss meant that NASA's aggressive launch schedule to complete the International Space Station was now rendered meaningless. He also realized that at that precise moment, the lives of the crew's families in the bleachers would enter an alternative future that he could not even begin to comprehend.

All he could manage to say aloud was, "This changes everything."

Standing next to O'Keefe, Bill Readdy was carrying a notebook that contained NASA's "Agency Contingency Action Plan for Space Flight Operations." Opening the notebook and reading the procedures from the start, Readdy told O'Keefe that he was declaring a spaceflight contingency. He officially activated NASA's Recovery Control Center at KSC.

Bridges urged O'Keefe, "Sir, we really need to go to an area where we can get our thoughts together on what to do next."

I was standing nearby. Still stunned, I told the VIPs to meet me in my office back at the Launch Control Center.

Ed Mango listened over the comm loop as LeRoy Cain instructed his mission controllers to lock the doors in Houston's Flight Control Room, the first step in impounding all the data. Mango activated a similar procedure in KSC's Firing Room. He instructed everyone to gather and record the data on their consoles, keep their logbooks at their desks, and not call anyone outside the room.

Staff in the Firing Room simply could not comprehend that the vehicle was gone. Feelings of shock and utter helplessness followed disbelief.

On the way back to my vehicle at the midfield park site, I phoned Mango and asked, "What do you know?"

He replied, "I don't think it's going to make it to the ground. I don't know what happened. They had some interesting data from the left wing that seemed to be getting worse, and then they lost comm."

I said, "The administrator will be there in ten minutes. He wants you to brief him on what you know."

I then tried to phone my wife Charlotte at home. She didn't answer. She was outside the house, hoping to hear the sonic booms and catch a glimpse from our yard of *Columbia* high overhead.

I left her a message: "*Columbia*'s not coming home. We don't know where it is. It's not here. I'll call you later."

# PART II

# COURAGE, COMPASSION, AND COMMITMENT

One sometimes contemplates their reason for being here on this earth or being involved in events of a specific place and time. Over the sixteen years I've lived in Hemphill, I've lamented over not living closer to my parents and have been frustrated with my lack of career advancement. I felt that God wanted me here, but I didn't really know why. The thought came to me that my role in this event might be the very reason that God placed and left me here in Hemphill.

—*Greg Cohrs, US Forest Service, June 2003*

# Chapter 5

# RECOVERY DAY 1

With the declaration of a spacecraft contingency at about 9:16 a.m. Eastern Time on February 1, what would have been Landing Day became Recovery Day 1. NASA immediately needed to determine precisely where *Columbia* was and ascertain the condition of the ship and her crew.

For many of us at NASA, and for the residents of East Texas, our lives had just changed forever.

## Accident Plus Twenty Minutes

In Tyler, Texas, Jeff Millslagle laced up his shoes for a training run for the upcoming Austin marathon. A rumbling sound startled him. As a California native, he at first thought it was an earthquake. But as the noise continued, he realized it was unlike anything he had ever heard.

Millslagle was one of the FBI's senior supervisory resident agents in Tyler. His colleague Peter Galbraith phoned him and asked, "What the hell was that?" They speculated that perhaps one of the pipelines running through the area had exploded. It seemed the only likely explanation. It was certainly not tornado weather. No other natural phenomenon could have caused such a prolonged banging.

Millslagle phoned the Smith County sheriff's office to see if they had any reports of unusual activity. They checked and phoned back,

"It was the space shuttle reentering." That didn't seem plausible, since the shuttle's sonic booms wouldn't be audible at sea level until the shuttle was well east of them. The sheriff's office called again a few minutes later. "The shuttle broke up overhead. There are reports that Lake Palestine is on fire."

Millslagle turned on his TV and saw video of *Columbia* disintegrating. He immediately phoned Galbraith and told him they needed to meet at the FBI office in Tyler.

He arrived at the office five minutes later, still dressed in his running clothes. Special Agent Terry Lane phoned in from his home west of Nacogdoches. Special Agent Glenn Martin called in from Lufkin, Texas. Both reported what appeared to be pieces of the shuttle on the ground. The sheriff's office phoned in and asked for guidance. Millslagle said, "Let me go home and throw some pants on, and I'll drive down to Lufkin." He called Lane and Martin and told them to meet him at the FBI office there. Galbraith offered to stay in the Tyler office and monitor the situation.

—

At Kennedy Space Center, the crowd at the Shuttle Landing Facility was struggling to comprehend why the space shuttle had not returned. Security personnel and the astronaut escorts quickly led the *Columbia* crew's immediate families away from viewing stands—and the eyes of the press—toward a special bus that would take them to the privacy and safety of the crew quarters.

The remaining guests, some of whom were members of the crew's extended families, were hurried onto buses and taken to the training auditorium in the industrial area of KSC about seven miles south of the landing facility.

Cell phones began ringing as the guest buses left the Shuttle Landing Facility. As word spread about what had happened, distressed passengers screamed and cried. Some demanded that the volunteer visitor escorts on the buses tell them what was happening. The escorts had no information to share. They were as confused and heartbroken as the passengers.[1]

Ann Micklos was waiting at the SLF with the landing convoy to greet her former boyfriend, *Columbia*'s Dave Brown, upon his return. She realized there was a serious problem when she saw the astronauts running toward the crew families' bleachers.

Ann immediately called Brown's parents from her cell phone. They were at home in Virginia watching the television coverage of the landing. The Browns were confused about the situation. They asked Ann if the shuttle was going around for another landing attempt. She explained that *Columbia* and her crew only had one chance to land. Her words caught in her throat as she told them it didn't look like they would be coming home.

As shocked and distressed as the spectators, the landing and recovery convoy teams at the runway briefly found themselves uncertain as to what they should do. The checklists that had taken them to this stage of operations were no longer valid. No procedures covered what to do in a scenario in which the recovery convoy was deployed but the shuttle did not come down at the landing site.

Reality began to sink in as the rest of the recovery team returned to their hangar. Workers were told not to talk to the press. They were also instructed to lock up all hardware and paperwork to impound everything for investigation.[2] They huddled in the hangar's foyer area. It seemed that everyone was crying and making phone calls to loved ones. Someone offered a few words of prayer.

Ann Micklos spoke up, sharing something that Dave Brown had told her before the mission. "Dave said, 'I want you to find the person that caused the accident and tell them I hold no animosity. I died doing what I loved.'"

Managers in the Firing Room started phoning supervisors, telling them to report to work immediately. Many of the people they tried calling were already getting ready to leave or were on their way to KSC. They had seen the stunning video on TV and wanted to help however they could.

—

At NASA's Dryden Flight Research Center on Edwards Air Force Base, California, Robert Hanley had just arrived at the standby landing team's trailer and heard them discussing Houston's inability to contact *Columbia*'s crew. He and the team watched the minutes tick by on the clock, without any updates from Houston. Growing nervous, Hanley called his colleague Judy Hooper in the crew quarters at KSC.

"Judy! Where's our crew?"

She replied, "I can't talk now," and hung up the phone.

Hanley instantly knew that something terrible had happened. Switching the television to CNN, he saw the pieces of *Columbia* fanning out like fireworks across the sky over Texas.

The room went silent.

Hanley looked at his team and said, "We lost the crew."

He held his emotions in check long enough to walk into the bathroom. Then he fell apart. *How could I lose my friends so close to home?* His thoughts went to the crew families at KSC, the people that he had come to know so well during the crew's training period. He normally would have been with them at the landing site. Now he was thousands of miles away, feeling helpless. There was absolutely nothing he could do.

The staff quietly cleaned up their paperwork. Hanley headed to the airport for what would feel like the longest trip home of his lifetime. Television monitors in the terminal at the Burbank airport replayed video of the accident again and again, while passengers went about their business in the terminal. Hanley wanted to stand on a chair and scream at everyone to shut up and think about what had just happened. He found a quiet spot where he could call his father. He cried with him on the phone until he regained his composure.

—

In Sabine County, Texas, Greg and Sandra Cohrs sat down to breakfast and turned on their television. Reports began coming in that

*Columbia* had "exploded" over Dallas. Grass fires were springing up in the area. Greg said to Sandra, "I bet we'll be involved in this before it's over." US Forest Service personnel, regardless of their job titles, typically were called in to help respond to all-risk or all-hazard incidents—wildfires, hurricanes, floods, and even terrorist attacks—in their local communities and across the nation.

Cohrs called the district fire management officer to find out if he should report for work. The officer said he was waiting for a call and told him to stand by. In the meantime, Cohrs prepared to do his usual Saturday yard work, but then a call back informed him that he would be on flight duty that day as a spotter.

Cohrs brought up the Intellicast weather radar website on his computer as part of his usual preflight routine. Despite the clear blue sky above, the radar image showed a wide swath of something in the air along a northwest to southeast track from Nacogdoches, Texas, through Hemphill and heading on toward Leesville, Louisiana. The largest concentration of radar returns was centered over Sabine County, and the cloud appeared to be slowly drifting north and east. He realized that the weather radar was picking up the debris from *Columbia* that was still falling to the ground. He took several screen snapshots of the radar display.

## Accident Plus Thirty Minutes

Sean O'Keefe and the other senior leaders huddled with me in my small conference room on the fourth floor of the Launch Control Center. We turned on a TV and saw for the first time the videos of *Columbia*'s plasma trail flashing and breaking up into smaller smoke trains.

Our hearts sank. We then knew for certain that *Columbia* was lost. There was no hope for her or the crew.

O'Keefe stared at the monitor. He put his hands on the table and said, "I wonder how many people on the ground we just hurt."

Associate Administrator Bill Readdy formally activated NASA's Contingency Action Plan for Space Flight Operations. Although it was rather general, the plan prescribed what NASA's leaders needed to do immediately to bring order to an emergency situation.[3] Among other things, the plan called for the formation of formal task forces to respond to and investigate the accident. Over the ensuing weeks, this list would grow to include one independent review board and fourteen formal internal task forces, working groups, and action teams.[4]

O'Keefe left the meeting and phoned President George W. Bush, who was at Camp David. Bush's first question was, "Where are the families?" O'Keefe was moved that the president's primary concern was to ensure that the families were being cared for. Bush then requested to speak with the families later that morning to express his personal remorse and to offer condolences from the nation. He and O'Keefe agreed that they would wait to place the call until the families had time to absorb the emotional blows of losing their loved ones.

While O'Keefe was out of the conference room, Roy Bridges asked me, "Mike, what do you think happened?"

I replied, "The only thing I can think of is the foam strike." I called for someone to bring us copies of the photos of the launch debris hitting the shuttle's wing.

O'Keefe returned, and I passed around the photos. I said, "We don't know if this is it, but there's nothing else about this flight that stands out. This is the only thing I can think of as a potential problem."

Everyone stared at the photos. Someone said, "Okay. Let's file this away for now. Don't jump to any conclusions."

We then discussed who should lead NASA's overall response to the accident and coordinate the various teams that would be responding from NASA's centers. Someone suggested appointing Dave King, who had spent nineteen years at KSC in roles including launch director, director of shuttle processing, and deputy center director. King had just recently moved to Marshall Space Flight Center in Huntsville, Alabama. O'Keefe and Readdy quickly agreed that King's experience made him an excellent choice to lead the recovery effort.

King, at home watching television, had just seen the news that NASA had lost communications with *Columbia*. Barely half an hour after the accident, his phone rang. Bill Readdy was on the line.

"Dave, do you know what's going on?"

"Well, I know something's going on, but I don't know what."

"I need you to run the recovery operation. I need you to go find our friends."

King instantly felt the heavy burden of the responsibility he was being asked to bear. Having been through the *Challenger* disaster and its aftermath early in his NASA career, he knew the things that could go wrong if the recovery of a ship and its crew was not handled properly.

King decided on the spot that he would do everything in his power to make this situation proceed better.

—

Five miles south of the Launch Control Center, Jerry Ross had raced back from the Shuttle Landing Facility to his office in the Operations and Checkout Building. He switched on the television to a news channel and saw the video of *Columbia*'s breakup.

He met the crew families as they arrived at the crew quarters on the building's third floor. He did not tell them what he had just seen on television. Rather, he did his best to make them as comfortable as possible.

STS-107 was Bob Cabana's first mission as the head of the Flight Crew Operations Directorate. He had ridden out with the crew on launch day, and he had been eagerly awaiting their return on the runway. Now he had to face the hardest thing he had ever done—telling the families that their loved ones weren't coming home. When he felt he had enough information, he and Ross joined the families in the crew conference room. Cabana explained to them that it was unlikely that any of *Columbia*'s crew could have survived the accident.

Ross spent the next several hours trying to comfort the families. He had their luggage collected from their hotel rooms, and he made arrangements to fly everyone back to Houston.

Ann Micklos returned to her office near the Vehicle Assembly Building after the landing convoy demobilized. Someone called to ask if she wanted to go to the crew quarters to be with the *Columbia* families. While she appreciated the gesture and sympathy about the loss of Dave Brown, she knew that her situation was very different.

"I chose to stay in the office and to try to figure out what happened with what data we had," she recalled. However, her coworkers could see that she was in shock.

After a while, they told her, "We're driving you home." She arrived to a house full of people to support her.

—

Things were going crazy in East Texas. In his office in Hemphill, Sheriff Maddox desperately needed to figure out what was happening in Sabine County. A deputy had just come on duty, and Maddox immediately dispatched him to the north end of the county. He phoned law officer Doug Hamilton from the US Forest Service and asked him to check on a possible train derailment near Bronson in the western side of the county. Maddox hopped in his car to head for the nearby natural gas pipeline. His dispatcher radioed him and said, "NASA just called and said it wasn't a pipeline explosion. That was just the shuttle going over and breaking the sound barrier. You can go about your regular duties."

Maddox drove over to Hemphill's youth arena to see what was going on at the livestock weigh-in. People asked him what it was that had passed overhead. Maddox told them that it was the shuttle breaking the sound barrier. One woman said, "But they haven't heard from it in fifteen minutes."

Maddox knew something was wrong. He got back in his car. The dispatcher radioed that people were calling in from all over the county about items raining down from the clear blue sky, breaking tree limbs, and hitting the ground.

Phone calls began pouring in to the town's volunteer fire department in addition to those coming in to the sheriff's office. "There are things falling out of the sky!" "Something just hit the road!" Then news of *Columbia*'s loss came through. The firefighters debated for several minutes about what to do. Volunteers started heading out to investigate the calls and secure the items being found.

Hemphill City Manager Don Iles arrived at the fire station as a radio report came in from a Texas Department of Public Safety (DPS) officer on US 96 near Bronson. "There's a big metal object in the middle of the highway. It gouged the road. I'm looking at it, but I don't know what to make of it."

The dispatcher asked if it had any identifying marks or numbers. The DPS trooper said, "Yeah, but it's only partial." He read back the numbers to the dispatcher, who then told him to wait. The trooper said he would stay at the scene by the object and keep vehicles from running over it.

A few minutes later, the dispatcher came back on and said, "I've just been in touch with NASA. Please do not pick up or touch any of the material, because it could be radioactive or poisonous." There was dead silence on the other end of the radio as the trooper pondered his situation.

Doug Hamilton arrived at the site of the reported train derailment, about five miles from his house, but there were no trains in sight anywhere. He called back to the sheriff's office and was asked to head to the wreckage sighting on US 96.

Sheriff Maddox and Hamilton met the DPS trooper at the scene. The metal object was the waste storage tank from *Columbia*'s crew module—the first confirmed piece of the shuttle found in Sabine County.

With the nature of the situation now confirmed, the accident scene in Sabine County officially became a federal incident. Hamilton, as a federal law enforcement officer, was now the man in charge.

Hamilton photographed the tank, and he and the others cordoned off the area with crime scene tape. They recorded its GPS position and called it back in to the dispatcher. Maddox and Hamilton had no officers available to guard the wreckage. The three men drove off together in Hamilton's government car to investigate the next reported sighting.

Someone overheard their radio report. When officials came back later in the day to retrieve the waste tank, it was gone.[5]

Hamilton, Maddox, and the trooper next responded to a call from a woman's farm outside Bronson. She had found partial human remains in her pasture. The sheriff went to the farmhouse and asked the family for a sheet to cover the remains. Someone suggested that they be removed from the pasture. Hamilton refused. "No, we ain't movin' nothin'! This is a crime scene." He now knew that the gravity of the situation required more than just taking photos and recording GPS locations. He photographed the remains and placed a sheriff's department officer on guard.

Hamilton and Maddox drove from scene to scene for the next several hours, meeting DPS troopers at the location of each reported sighting. A few findings appeared to be partial remains of the crew, but most were pieces of the shuttle.

At one house, an aluminum I-beam had fallen through the carport roof, broken through the concrete floor, and buried itself in the ground. This eventually turned out to be the only structural damage sustained anywhere in Sabine County.

At that time, Hamilton was the only law officer in Sabine County who had a digital camera. After visiting the first six or seven scenes, Hamilton had already filled two data disks with photos. He realized it would be physically impossible for him to visit all the debris sightings, especially now that he knew bodies were on the ground. And there were not enough law officers in Sabine County to guard every piece of debris being discovered.

## Accident Plus One Hour

Astronauts Mark Kelly and Jim Wetherbee were at their suburban Houston homes when contact with *Columbia* was lost. Both immediately knew the situation was dire. They rushed to the astronaut office in Building 4 South at Johnson Space Center.

As quintessential type A personalities, astronauts are biased toward acting to bring a situation under control. Patience can be a tough virtue for them to exercise, especially when the lives of their colleagues are on the line. While they awaited official orders, the astronauts at JSC took whatever actions they could. Mark Kelly and rookie Mike Good brought out the contingency checklist and reviewed the required actions. Working through the list, Kelly and Good made phone calls, and within fifteen minutes, nearly fifty people were in the astronaut office conference room. Discussions began on how to farm out the astronauts to locate *Columbia*'s crew.

Several of the astronauts decided to head to Ellington Field, about halfway between JSC and Houston. Ellington was the home base for the T-38 jets the astronauts flew around the country. Wetherbee drove home to pick up his flight suit and pack his overnight bag, and then drove to Ellington to await orders.

Kelly pointed out to Andy Thomas that the contingency plans never envisioned the shuttle coming down within a two-hour drive of Houston. Kelly said, "We really need to send somebody to the scene right now."

Thomas said, "Okay. You go."

Kelly considered his options for getting north to the accident scene as quickly as possible. He phoned Harris County constable Bill Bailey and requested a helicopter. Bailey made a few calls and phoned back. "I'm sending a car to pick you up. The coast guard is going to take you up there." Kelly grabbed astronaut Greg "Ray J" Johnson to accompany him. They arrived at Ellington Field and boarded the waiting helicopter.

As they took off, the pilot asked, "Where are we going?"

Kelly said, "I heard there's debris coming down at Nacogdoches. Let's go to the airport there."

—

While the astronaut corps mobilized, NASA management activated its Mishap Investigation Team (MIT) in the Mission Control Center at JSC. NASA appointed the MIT members as a routine matter prior to each space shuttle mission in case anything were to go wrong during the mission. Dave Whittle had been identified as the MIT chairman for STS-107.[6]

Whittle was the chairman of NASA's system safety review panel and the safety manager for the Shuttle Program. He was certified as an aircraft accident investigator by the National Transportation Safety Board and the University of Southern California. He also had extensive experience investigating space accidents.

Whittle usually attended shuttle launches, but was off duty for most landings. He was at home when Mission Management Team chairperson Linda Ham called to say, "We think we're going to need you."

He was unaware at that point that *Columbia* was lost, but he knew something bad must have happened for Ham to call him at home on a Saturday morning. He told his wife, "Pack my stuff. I may be gone a long, long time."

Astronaut Dom Gorie had just returned home to Houston that morning from a vacation in Hawaii. While he and his wife were unpacking, someone called and told him to turn on his television. Gorie was the designated astronaut representative for Whittle's investigation team. He realized that he had to get to the office as quickly as possible. He grabbed his pre-packed "deployment" bag, said good-bye to his wife, and was at JSC to assist Whittle in less than an hour.

Whittle did not know the full extent of the situation until he was briefed in the Mission Control Center, where the Mission Management Team was gathering. After a quick briefing, Space Shuttle Program Manager Ron Dittemore and Ham told Whittle, "You've got about thirty minutes to come back in here and tell us what you want to do."[7]

Whittle's team would be among NASA's first responders at the scene. In this circumstance, he would be in charge of collecting and protecting all shuttle debris and impounding it for the eventual official investigation of the accident. His scope also included the sensitive matter of handling the remains of *Columbia*'s crew. He needed to set up a command center. It had to be a secure location, away from the press and other prying eyes, able to accommodate multiple types of support aircraft from cargo planes to T-38 jets, and had to have appropriate support facilities such as a morgue and a staging warehouse. If he needed to set up a headquarters in the field, the Department of Defense Manned Spaceflight support group was available to provide for NASA's infrastructure needs, with tents, helicopters, food, and other necessities. However, Whittle thought it might be best to manage the investigation from a military base somewhere near the shuttle's last known location. *But where was that?*

Reports were coming in of debris hitting the ground from near Dallas to Fort Polk, Louisiana. Looking at a map, Whittle saw that Carswell Naval Air Station near Fort Worth was too far north and west of the reported sightings. Fort Polk seemed too far south. Barksdale Air Force Base, near Shreveport, Louisiana, looked like a possible option for the MIT's strategic command center.

Whittle called the vice commander of the Second Bomb Wing at Barksdale, Colonel Charles McGuirk. The base and its contingent of B-52 bombers were on alert for possible imminent action in Iraq. Despite the tense military situation, McGuirk immediately offered to host the MIT at the base. The facilities and computers from a court-martial hearing at the base were still in place and available for NASA's immediate use. Whittle and his team could fly in and be in business that day.

—

While Dave Whittle was getting his team ready for action in Texas, NASA's senior officials who were still at the Launch Control Center at Kennedy had to initiate a broader investigation.

One requirement of NASA's agency-wide contingency plan for major incidents was to set up an independent review board—one not under NASA's direction. Bill Readdy called the people who were named in the plan as members of the accident investigation board. These individuals had the requisite technical, scientific, and organizational expertise to serve on the panel.

The contingency plan did not name the board chairman. That was left to the administrator's discretion, driven by the nature of the circumstance.

Sean O'Keefe's thoughts turned to Admiral Harold "Hal" Gehman Jr., who had a background in complex bureaucratic organizations and had just completed an investigation of the terrorist attack on the USS *Cole*. Gehman's temperament and experience seemed perfectly suited to lead the *Columbia* accident investigation.

Gehman heard about the accident just before O'Keefe's deputy administrator Fred Gregory phoned to ask him to lead the board.

**Accident Plus Ninety Minutes**

As he left Tyler in his car and drove toward Lufkin, the FBI's Jeff Millslagle recalled other incidents he had investigated, including the 1996 crash of TWA Flight 800, which left parts of the plane and remains of the passengers floating in Long Island Sound. He thought that by comparison, this would be a small incident. *After all, how much of a shuttle could possibly survive a breakup and reentry from so high up?*

He was twenty minutes out of Tyler when he saw a Texas Department of Public Safety officer standing by a hole in the mud beside the roadway. Millslagle asked the trooper what he was doing. "This is a piece of the shuttle. We've been instructed to guard these things."

The closer Millslagle got to Lufkin on his ninety-mile drive, the more debris he saw.

—

Jan Amen, assistant to the state fire chief in the Texas Forest Service, had heard the explosions at her home in Nacogdoches County. She drove to the Etoile Fire Department, where she was a volunteer firefighter. She encountered pieces of the shuttle along the way. Many were like one she saw lying in the middle of Highway 103, which was over one foot in length, had three burn holes through the middle, and was twisted into a bowl shape. She and her fellow firefighters flagged and stood by pieces of debris while waiting for the Department of Public Safety to arrive.[8]

—

One hundred and fifty miles south of Etoile, NASA's Mission Management Team gathered in the Action Center in the Mission Control Center building about an hour after the contingency was declared. Personnel from Kennedy, including my management team and me, teleconferenced in from a large room on the first floor of the Launch Control Center. Representatives from all the major NASA and contractor organizations either joined in person or phoned in. Ron Dittemore led the meeting. The atmosphere was solemn. As people took their places, they speculated on what might have caused the accident.

O'Keefe promised NASA's full support, and he delivered moving words of sorrow and resolve that deeply touched the team. He announced that he had activated the International Space Station and Space Shuttle Mishap Interagency Investigation Board (which shortly thereafter was renamed the *Columbia* Accident Investigation Board [CAIB]), with Admiral Hal Gehman as its chairman.[9]

Dittemore said he was receiving reports that some debris was on the ground in Texas and Louisiana. The first question asked was, "How hazardous is it?" No one knew the extent to which the toxic and explosive materials aboard *Columbia* might survive reentry and make it to the ground.

Managers began to focus on recovery.[10] Dave Whittle and his Mishap Investigation Team would lead NASA's internal investigation by recovering the physical debris. Dittemore's voice broke with emotion as he said, "Let's go get the crew. We can't leave them out there."

The question arose, "What can Kennedy do to help?" Ed Mango and I huddled briefly with our team. After a few moments, someone suggested, "Why don't we treat this like a TAL landing?"

One of the shuttle's launch abort modes, in case an emergency prevented the vehicle from reaching orbit, was a transoceanic abort landing (TAL). Every KSC contingency plan assumed the shuttle would come down within sight of a runway or make an emergency landing at one of the TAL sites in Europe or Africa. In the event of a "nonroutine" landing, Kennedy was to deploy its Rapid Response Team (RRT), about eighty KSC engineers and technicians who were experts on the shuttle and its systems. Their role was to get the situation under control, retrieve the shuttle, and bring it back.[11] The RRT seemed like the appropriate group to support this incident, even though this was unlike anything we had ever envisioned.

A few members of the RRT were trained in how to conduct crash investigations. However, we had never trained for a scenario where the orbiter broke up far from the landing site. Such a situation had always been considered a "non-credible event"—something too unlikely to happen. The emergency procedures would not be much help, either. No one yet knew where—or even if— there would be a shuttle to retrieve. NASA had no reliable information yet regarding if or where *Columbia* had fallen to the ground or if it had splashed into the Gulf of Mexico. We only knew that the RRT would deploy somewhere between Dallas and Florida.

I formally suggested that NASA send the Rapid Response Team to support the recovery. Dittemore and the senior leaders liked the idea. Leadership of the RRT normally fell to Kennedy's processing flow director for the mission—Scott Thurston, in this case. However, everyone felt that a more senior person needed to head the team in this complex situation.

They appointed me to command KSC's forces in the field.

Dittemore asked me, "When can you deploy?" I said that the air force always kept a transport plane on reserve for NASA in case of launch emergencies. It was not stationed at KSC, but could be brought to the center within six hours.

Immediately after the meeting, my team and I discussed how to begin the investigation and recovery process. We started identifying names and thinking about the logistics of getting those people onto planes to Texas as quickly as possible, either on the first transport flight that afternoon or the next day. Managers tracked people down at their children's soccer games or other weekend activities and told them to pack a bag and get to KSC as quickly as possible. While no one knew yet where the team would deploy, there was no time to lose while waiting for better information. It was imperative to get people moving immediately to start securing any hazardous debris.

One of the people contacted was Linda Moynihan, who provided administrative support for United Space Alliance's director of safety and quality. She was told to pack a bag for what she thought would be a week's stay in Texas. Then she visited as many automated teller machines as she could to withdraw cash for the team in the field. Arriving in her office at KSC, she filled two copy-paper boxes with office supplies.

I called my wife Charlotte and asked her to check my "TAL packing list," pack my suitcase, and bring it to Kennedy. I saw her shortly after lunchtime. I hugged her tight, not knowing when I would see her again.

## Accident Plus Two Hours

Billy Ted Smith was the emergency management coordinator for the East Texas Mutual Aid Association, which included Jasper, Sabine, and Newton Counties. He phoned Sheriff Maddox to say that he was on his way to Hemphill with Jasper sheriff Billy Rowles, Jasper police chief Mark Allen, and several other men. They joined Maddox, Doug Hamilton, Olen Bean of the Texas Forest Service, Sabine County Judge Jack Leath, and representatives of Jasper, Newton, and Sabine Counties in Hemphill to set up an incident command post.

Maddox initially considered locating the command post at the Bronson fire hall, close to where the first partial crew remains had been found. However, Hemphill offered somewhat better communications, although the town's facilities would likely still be inadequate.

Hemphill was a rural town of about eleven hundred residents in a county with a total population of about ten thousand souls. Its infrastructure would shortly be stretched to the limit. In early 2003, the community had only two high-speed T1 Internet lines. Maddox ordered two drops to be run from the high school's T1 line to the fire station. Only one phone line ran into the firehouse. Maddox called the phone company and ordered ten lines to be installed immediately. Before the installation was complete, he requested that the number of lines be increased to twenty—then to thirty.

Maddox ordered all of Sabine County's emergency personnel to report for duty. Firemen removed trucks and equipment from the fire station bays and parked everything across the street. People went to work setting up the command post. Two inmates in the town jail were carpenters. Maddox directed them to pick up plywood and two-by-fours and start building work cubicles.

Cecil Paul Mott, Hemphill's electrical supervisor, was called in to upgrade the electrical service to the fire hall. After installing outlets, he had to replace the transformer outside the firehouse. By the time the week was over, he had filled all the power poles in the town center to capacity. He resorted to hanging transformers and lines from trees.

Fifty miles to the west, Jeff Millslagle arrived at Lufkin's FBI office shortly after ten o'clock. The assistant US attorney was on hand, as were all of the area's FBI agents. Phones were ringing incessantly. The agents discussed what the FBI's role should be. Their first concerns were to determine whether anyone on the ground was injured by falling debris and whether the shuttle had been shot down or sabotaged.

Millslagle sent agents Terry Lane and Shane Ball to Hemphill and dispatched Glenn Martin to San Augustine. Most of the debris calls were coming from those two areas.

US marshals and representatives from the Bureau of Alcohol, Tobacco, and Firearms arrived at the office, as did more people from the US Attorneys Office. It was clear that the FBI's office was too small to serve as the central coordinating location. Brit Featherstone from the US Attorneys Office made several calls and secured the Lufkin Civic Center for use as an incident command post.

Back at the FBI office in Tyler, Pete Galbraith was having difficulty getting the attention of his superiors and conveying the urgent need to deploy more FBI support to the scene. "What's this got to do with counterterrorism?" was their initial response. They seemed unable to grasp the severity of the problem.

Millslagle phoned his superiors in Denton and received similar pushback. Millslagle said, "I don't think you guys get how big this thing is about to become."

Meanwhile, Jack Colley, the head of the Texas Department of Public Safety's Division of Emergency Management, phoned Mark Stanford, chief of fire operations for the Texas Forest Service. Stanford had extensive experience in implementing the incident command system (ICS) to manage large all-hazard incidents. While the Texas Forest Service might lack the technical expertise to deal with the accident itself, the incident command framework provided a support structure for managing the response to a complex public safety situation. Colley knew of Stanford's experience and asked him to get to Lufkin as quickly as possible to take control and coordinate the agencies that were responding to the accident.

## Accident Plus Two Hours and Thirty Minutes

Scott Wells was deployed in Jonesboro, Arkansas, as the Federal Emergency Management Agency (FEMA) on-site incident commander. Wells was a military veteran with twenty-four years of service. He left the military in 1999 to become part of FEMA's original cadre of federal coordinating officers. On February 1, he was helping the

Jonesboro community recover from power outages resulting from a severe ice storm earlier in the winter.

He received a call from his regional administrator, Ron Castleman, informing him that there had been a disaster involving the space shuttle. Castleman had few details to share, but he told Wells, "Be prepared to leave shortly."

Half an hour later, he phoned Wells back again. "Start heading toward Barksdale Air Force Base. There will be more information to follow."

President Bush had declared a state of emergency in Texas and Louisiana. He authorized FEMA to coordinate and direct all other federal agencies that might be responding to the accident. Castleman assigned Wells as FEMA's leader on the scene.

During the six-hour drive southwest to Barksdale, Wells teleconferenced with Castleman, FEMA headquarters in Washington, NASA headquarters, and Tom Ridge, the head of the new Department of Homeland Security. Priorities for NASA and FEMA were clear. First and foremost, they were to ensure public safety. Recovery of the crew's remains was the next priority, followed by supporting NASA's accident investigation.[12]

Wells also received a call from Jack Colley at the Texas Department of Public Safety. Colley gave him the first detailed situational awareness of the conditions on the ground in Texas, because he had been in contact with the National Guard and Texas Department of Public Safety people on the scene. Colley's two biggest fears were that this was somehow linked to a terrorist attack, and that the hazardous materials on the shuttle could endanger public safety.

—

Astronauts Mark Kelly and "Ray J" Johnson arrived at Nacogdoches airport at midmorning in their coast guard helicopter. They learned that debris had been coming down all around the area, and that some had even hit the ground at the airport.

The astronauts walked to a hangar to examine some of the pieces that had been collected. Kelly instantly recognized a fuel tank from the shuttle's maneuvering engines. Alarmed, he told the workers in the hangar not to go near the tank, as it might be contaminated by highly toxic propellants.

Kelly and Johnson boarded a police car to ride around town and assess the situation. Wreckage from the shuttle appeared to be strewn everywhere—even along the roadway. Kelly wondered: *How could you even begin to get control of this situation?*

The astronauts returned to the car after examining a piece of debris on the roadside. One of the police officers said, "We're getting reports that some of the crew might have come down near Hemphill." Kelly and Johnson knew that they needed to get there as quickly as possible. They returned to the airport and boarded their helicopter, directing the pilot to take them to Hemphill, sixty miles to the southeast.

—

Administrator Sean O'Keefe and Associate Administrator Bill Readdy met with *Columbia*'s families in the Kennedy Space Center astronaut quarters at 11:30 a.m. Eastern Time. O'Keefe told the families that President Bush wished to speak with them by conference call. The president conveyed his deepest personal regrets and offered his full and immediate support for whatever actions needed to be taken.

The group spent the next ninety minutes working through details of the next steps. During the conversation, one of the spouses said the most important thing NASA could do was find out what happened, fix it, and rededicate the agency with everything possible to achieve the exploration goals that the crew gave their lives for. O'Keefe was heartened by their encouragement at such a dark time.

## Accident Plus Three Hours

Astronaut Dom Gorie made calls to find out if any reconnaissance aircraft were available that could aid in the search for *Columbia* and her crew. He learned that the Drug Enforcement Administration had a Fairchild C-26 airplane stationed at Ellington Field. The C-26 was equipped with forward-looking infrared equipment. If any large pieces of still-warm debris from the shuttle were on the ground, the sensors might be able to detect them.[13]

Gorie secured the use of the plane and flew with them to direct the airborne search. No shuttle material was found during their three-hour flight. Disappointingly, the infrared imaging system could not resolve any small pieces of debris on the ground, because large areas of still-smoldering tree stumps from recent controlled burns looked surprisingly like a debris field of warm objects to the infrared sensors.

—

Dave King drove to Marshall Space Flight Center in Huntsville shortly after Bill Readdy asked him to head NASA's overall recovery effort. A group briefed him on the current situation. Reports of possible debris sightings were coming in from California to Florida.

King asked that Marshall's plane be readied to take him to the accident scene. He phoned his security team and his information technology department and asked that they designate personnel to accompany him. Their departure time would be "as soon as the pilots are ready."

King drove to Huntsville airport to meet with his team. They quickly discussed what they needed to accomplish once they arrived at their destination—wherever that would be. Several options were being considered in East Texas. The decision would be made while King was en route from Huntsville toward Texas.

Their plane took off at about 11:30 a.m. Central Time and headed southwest. While the plane was in the air, King learned that FEMA was going to set up its disaster field office in Lufkin, Texas. King told the pilot to set course for Lufkin.

—

Mark Kelly's coast guard helicopter set down on Hemphill high school's football field. Kelly went into the school gym, where a basketball game was in progress. He found a policeman and asked to be taken to the town's incident command center. The policeman escorted him to the firehouse, about one-quarter mile south. Kelly introduced himself to the FBI's Terry Lane, who had also just arrived in town.

About three hours after the accident, a call had come in regarding a sighting of something unusual on Beckcom Road, a few miles southwest of town. A jogger had seen what he first thought to be the body of a deer or wild boar near the roadway.

Kelly and Lane rode with Sheriff Maddox to the site. They met Tommy Scales from the Department of Public Safety, who had just come from another debris scene nearby.

They encountered what was clearly the remains of one of *Columbia*'s crew.[14] Maddox radioed John "Squeaky" Starr, the local funeral director, to come to the scene to assist in the recovery. Kelly also requested that a clergyman come to the site to perform a service before the remains were moved or photographed. While they were waiting, another state trooper covered the crew member with his raincoat.

"Brother Fred" Raney, the pastor at Hemphill's First Baptist Church, had just returned to the firehouse to report that he had seen partial remains of a *Columbia* crew member in a pasture. Raney drove out to join the officials gathered at the Beckcom Road site. He conducted a short memorial service for the fallen astronaut.

By now, the news media had arrived in the area and were monitoring police radio frequencies. They intercepted the call about the Beckcom Road remains, and a news helicopter full of reporters flew over the scene, trying to get video of the recovery on the ground. They were low enough, and their motion deliberate enough, that it was clear to the people on the ground that the pilot was trying to use the helicopter's rotor wash to blow the raincoat from the crew member's remains. Lane and two troopers stood on the corners of the coat to keep it in place. One of the troopers used "an emphatic gesture" to make it clear to the pilot that the helicopter had to leave the area immediately.

Lane later said, "I don't know if a helicopter has ever been shot down by a DPS pistol, but they were very close to that happening."

The troopers noted and reported the tail number of the helicopter. Shortly thereafter, the FAA ordered a temporary flight restriction over all of East Texas.

After the remains were placed in a hearse, Kelly, Lane, and Raney moved on to a house near Bronson, where partial remains of another crew member had been reported. The media had intercepted those radio calls, too, and several reporters were on the scene when the officials arrived. From that point forward, the command team began using code words and decoy vehicles whenever crew remains were being investigated.

Lane was a veteran of the FBI's response to many horrific accidents, including the TWA 800 crash and the World Trade Center attacks. His orders in East Texas were to work as part of a team of four individuals: himself, a forensic anthropologist, a pathologist from El Paso, and a Texas Ranger. Whenever a call came saying, "We think we've found something," the team's role was to ascertain how likely it was that the finding actually was human remains. The team deployed all necessary resources to make a recovery of any remains that were probably or likely to be human.

Making the first two crew remains recoveries in the space of a few hours set the tone and protocol for subsequent recoveries. From that point forward, whenever possible crew remains were located, the FBI would be called to the scene immediately. An astronaut accompanied Lane and his team to investigate every sighting, without exception.

A Texas Ranger or DPS officer, Brother Fred, and a funeral director (either Squeaky Starr or his son Byron) would meet the FBI Evidence Response Team and astronaut at the site. Once the group was assembled, the scene was turned over to the astronaut to step forward and preside over the recovery of his or her colleague. When the astronaut was ready, he or she would signal Brother Fred to approach, who then performed a brief service offering a few words on the heroic sacrifice of the crew member, reading certain verses from Scripture, and then saying a prayer. The astronaut then released control of the scene to the FBI Evidence Response Team. The remains were placed in a body bag and taken to a hearse. Squeaky or Byron Starr would then transport the crew member's remains to a waiting doctor.

Brother Fred learned that *Columbia*'s commander Rick Husband had recited Joshua 1:9 to his crew as they suited up before their flight: "Have I not commanded you? Be strong and of good courage; do not be afraid, nor be dismayed, for the Lord your God is with you wherever you go." Brother Fred incorporated that verse into all of his services in the field. He and Kelly also researched appropriate words to say for Hindi and Hebrew services. All of the services delivered in the field after the first day included Christian, Jewish, and Hindu prayers.

Lane was deeply moved by the way the astronauts handled the recoveries. He knew how horribly difficult it must have been for them to see their friends and colleagues in that condition. It was a duty far outside the scope of what any astronaut would normally be asked to perform. The strength of the astronauts' religious convictions also surprised Lane at first. Then he realized that if these people "strap a million pounds of dynamite to their butts for someone else to light, they'd better have mighty deep faith."

**Accident Plus Three Hours Thirty Minutes**

Greg Cohrs was still at his home near Hemphill when he received a call from district ranger Marcus Beard, directing him to report to the US Forest Service office, don his flight suit, and begin searching the area in a government-contracted helicopter. Cohrs arrived at the office only to learn that the helicopter was grounded because airspace over the entire area had just been restricted. Beard then sent Cohrs to the Hemphill fire station to attempt to establish joint command of the incident in Sabine County with Sheriff Maddox.[15]

The firehouse was a maelstrom of activity. Maddox was too busy to talk. Cohrs found Billy Ted Smith—the co-incident commander with Maddox—and offered the services and resources of the US Forest Service, which administers one-third of the land in Sabine County. Cohrs and his colleagues knew the deep woods and terrain as well as anyone. They were experienced in search and rescue methods, and they had heavy equipment to help clear paths into the woods.

Reports and unconfirmed rumors were coming in from other locations. One said that the Department of Defense would shortly take over the search-and-recovery operations. Another said that the army was being deployed. There might be hazardous or radioactive material in the field. There was classified material on the shuttle. A storage tank on the ground in San Augustine County was venting yellowish gas. Solid intelligence was still hard to come by. What was clear was that all shuttle materials, except personal items and avionics, were to be protected and left in place for subsequent retrieval. Crew remains needed to be recovered immediately.

Cohrs immediately began to organize search efforts in Sabine County. He divided his Forest Service personnel into two-person teams and began to assign priorities to investigate the reported findings. He assigned highest priority to sightings of possible crew remains. He instructed the teams to note the GPS coordinates of debris finds and log them for later recovery.

Meanwhile, representatives of various state and local agencies, who had been ordered by their leaders to help out in Sabine County, began to overrun Hemphill's town center. Each agency tried to stake out space in the firehouse. Other volunteers were arriving faster than they could be handled.

Media trucks were descending on the town, occupying valuable space along roads that would be needed by emergency response teams. Sabine County Judge Leath ordered the area around the courthouse to be cordoned off.

Sheriff Maddox called Roger Gay, commander of Hemphill's VFW Post, which was four miles from the town center. "Roger, you need to open up the VFW hall. We need to use it as a staging area."

Gay replied, "No problem."

Maddox added, "Could you come up with some sandwiches and coffee? Maybe make some tea?"

Gay again replied, "No problem. We'll get it going."

Gay went to the VFW hall and started making sandwiches. At lunchtime, a few people came in for food. Then a few more arrived, and then a few more—and then dozens more.

Realizing that he would quickly be overwhelmed, Roger phoned his wife Belinda, who was on her way to a baby shower in Nacogdoches. He told her he needed her help as head of the VFW Ladies' Auxiliary. She made a U-turn and sped to the VFW hall. They began making calls, asking people to prepare food and bring it to the VFW hall.

**Accident Plus Four Hours**

NASA's Mission Management Team held a second meeting at JSC to report status and update plans. The Department of Defense, the Southeast Air Defense Sector, and the National Transportation Safety Board were tied in and offered their support. FEMA and the coast guard were online. The Department of Homeland Security was involved, along with the State Department. The Rapid Response Team and I teleconferenced in from Kennedy.

Reports indicated that debris was being found from Tyler, Texas, to Louisiana. No signals from the emergency beacons on the crew's suits had been detected. It was clear there was no hope of finding the crew alive. The astronaut office reported on how the crew's families were being cared for. Discussions began about a memorial service.

My Rapid Response Team was scheduled to board an air force cargo plane bound for Barksdale Air Force Base at about four o'clock that afternoon. The plane could carry ninety-six passengers. Once we arrived, we would report to Dave Whittle to support his Mishap Investigation Team. Whittle would be reporting to Dave King.

**Accident Plus Five Hours**

In the early afternoon, Jim Wetherbee and several other astronauts began driving north from Houston to Lufkin. Along the way, they received reports of shuttle equipment that had been found on the ground.

Wetherbee was directed to one site to check out a sighting of particular importance. There he saw the charred and damaged remnants of a shuttle astronaut's helmet in a field. He and his team made sure that the scene was secured so that the equipment would be untouched until it could be recovered properly. The helmet was a harsh reminder of his seven friends on board *Columbia* and what they must have gone through when their vehicle came apart.

Meanwhile, the majority of Dave Whittle's Mishap Investigation Team had assembled in Building 30 at Johnson Space Center. The team's two designated flight surgeons would normally have been present, but they were at Kennedy with the crew's families. In their place, NASA flight surgeon Philip Stepaniak, MD, and Michael Chandler of Wyle Laboratories were assigned as the medical representatives to the MIT. They would be responsible for transferring any recovered crew's remains to the Armed Services Institute of Pathology at Dover AFB for autopsy.

At the first official meeting of his team, Whittle announced that he and select members of the MIT would be deploying to Barksdale Air Force Base in Louisiana that afternoon to set up a strategic command center. Whittle learned that FEMA was also being activated and would be setting up their command center in Lufkin because of the town's proximity to the debris field. Being situated in Lufkin facilitated FEMA's ability to get resources to the local communities who needed help. Whittle decided to situate leaders from his NASA team at both Barksdale and Lufkin.

—

Recovered shuttle debris was already appearing for sale on eBay. Jeff Millslagle asked the Houston FBI office to shut down those listings immediately. Shuttle material was government property, and unauthorized possession was a federal crime.

Dave King's plane from Huntsville landed at the Angelina County Airport outside Lufkin. The sheriff, an FBI special agent, and a Secret Service agent met him and gave him a brief situation report.

King arrived at the FBI office shortly after one o'clock. Jeff Millslagle took him aside and found him to be still somewhat in disbelief about the situation. "We never anticipated this happening on reentry," King told him. "The shuttle's just a flying brick."

King needed to start making decisions immediately about recovery operations. He had never run a recovery operation like this before. But then again, nobody had ever done anything like this.

Millslagle asked what help NASA needed from the FBI. King explained that the aftermath of the *Challenger* disaster had been a logistical mess. King did not want a repeat of that nightmare.

"I'd like the FBI crime scene folks to work along with locals and astronauts, when we start searching for human remains and some of the sensitive equipment," he said, "to preserve those remains and items. No pictures. Nothing like that. We don't need it on the news. But we have to do it properly so that we can get the crew's remains back, if there are any."

Millslagle replied, "You got it."

—

President Bush addressed the nation from the White House Cabinet Room at 2:04 p.m. Eastern Time. "My fellow Americans," he began, "this day has brought terrible news and great sadness to our country. At nine a.m., Mission Control in Houston lost contact with our space shuttle *Columbia.* A short time later, debris was seen falling from the skies above Texas . . .

"The *Columbia* is lost. There are no survivors."

## Accident Plus Six Hours

Chief Flight Director Milt Heflin and Space Shuttle Program Manager Ron Dittemore said in a press conference at Johnson Space Center that it was too early to blame the accident on the foam strike on *Columbia*'s wing. Dittemore noted that the Mission Management Team had concluded while *Columbia* was in orbit that the foam strike was not a concern. He reiterated, "That is the case today. We have no information that would say that is not the case."[16]

—

Calls were flooding into the Sabine County command center at an alarming rate. Maintaining order was nearly impossible. Teams that had been sent out to the field to investigate calls were not returning to the command center; they were constantly being interrupted or distracted by worried citizens. As soon as responders left a scene after investigating one sighting, someone else would flag them down along the road. Greg Cohrs eventually had to rein in the teams and instruct them to investigate only their assigned sightings. If stopped by citizens in the field, they could take information and report it to the dispatcher to be dealt with sequentially, but they could not stay to investigate further. It was the only way to regain control on the first day of the recovery.[17]

Searches in the field that afternoon were loosely structured. Searchers were looking for anything of interest, primarily any possible remains of the crew or their personal effects. Many of the search teams did not have GPS equipment. They called in descriptions of where items were found and flagged them to be retrieved later.

Shuttle debris was everywhere around them.

—

A half hour later, Dave Whittle called Ellington Field to arrange transportation for the Mishap Investigation Team to Barksdale. NASA's infamous *Vomit Comet*—the Boeing 707 used for zero-G training flights—flew the JSC contingent and the MIT's equipment to Barksdale. They departed Ellington at about 3:30 p.m. Central Time.

At Kennedy, the Rapid Response Team and I should have been leaving at about the same time. However, there were delays getting an air force plane to us. It was frustrating to wait in the Launch Control Center when we were so anxious to get to the scene and start working.

**Accident Plus Eight Hours**

Jim Wetherbee and his carload of astronauts arrived at Lufkin in the late afternoon. Wetherbee found the FBI's Jeff Millslagle and introduced himself as the representative from the astronaut office.

Wetherbee asked, "How can I help?"

Millslagle looked at him. "What's your plan?"

Wetherbee thought that perhaps he hadn't spoken clearly, and he repeated, "I'm Jim Wetherbee from the crew office in Houston, and I'm here to help."

Millslagle once again asked, "What's your plan?"

Wetherbee then began to realize the responsibility he bore as the senior astronaut representative. People were looking to *him* for direction and answers. He knew he had to be quick and accurate in establishing a plan of action.

Wetherbee located Dave King, and the two discussed their roles. King would be the "up and out" person, interfacing with the senior leaders of the many agencies involved, the White House, and NASA headquarters. King assigned Wetherbee as the crew recovery's "down and in" leader, with operational decision authority for anything related to searching for and recovering *Columbia*'s astronauts. He would interface with the local FBI, FEMA, the Environmental Protection Agency (EPA), the Texas Army National Guard, the Texas Department of Public Safety—and what eventually grew to be forty-four different federal, state, and local organizations involved in the search for the crew.

Wetherbee took a deep breath and got to work.

The NASA checklists and contingency plans at Wetherbee's disposal contained valuable information—names, phone numbers, and roles and responsibilities. All this would come in handy later on in the investigation. However, it was inadequate for the immediate situation, which was chaotic and still evolving.

He drew on his naval experience and training for what to do in a catastrophe like the one he was facing. First, he would establish his command center, find out what resources were available, and gather intelligence from the field. Then, he could develop a plan and staff the key positions with his fellow astronauts. Only then would he feel that he had taken control of the situation rather than reacting to it.

FEMA's Emergency Operations Vehicle arrived in Lufkin from Denton, Texas, late in the afternoon. FEMA parked the eighty-two-foot-long tractor trailer in the lot adjacent to the Civic Center. A section of the trailer expanded sideways, forming a work area that accommodated a twenty-five-person FEMA response team. Equipped with electrical generators, satellite radio transceivers, computers, desks, file servers, printers, and copiers, it was FEMA's self-contained command center.[18]

Meanwhile, personnel from FBI, NASA, and other agencies moved from Lufkin's FBI office to the Civic Center and began to set up shop.

Mark Stanford arrived from the Texas Forest Service office with a truckload of supplies. He went inside the Civic Center to take a look at the situation. To his trained eye, the command center was beginning to take shape, but it would have appeared as complete chaos to an outsider. Representatives from myriad state, local, and federal agencies were scurrying around and carving out space for their operations. Stanford understood that this was completely normal, given the magnitude and recency of the tragedy and the number of agencies involved.

Stanford knew that he would be immediately shot down, laughed at, or ignored if he announced that "the Texas Forest Service is here to bring order to the situation!" His experience showed that the best way to get a team to embrace you in a disaster situation was to make yourself as helpful as possible. "Figure out who's in charge by observing the scene. Then go up to that person and say, 'What three things are biting you on the ass?' And then you make it your goal to fix those three things. Then you're part of the team."

**Accident Plus Eight Hours Thirty Minutes**

Upon Dave Whittle's arrival in Barksdale Air Force Base at 4:30 p.m., the base's deputy commander introduced himself and took Whittle to the facilities assigned to NASA. He again promised the military's full support for whatever the Mishap Investigation Team needed—office space, lodging, transportation, and so forth.

Now it was time to establish Whittle's strategic command center at the base. He divided up the team into their areas of responsibility. By 5:30, the team's flight surgeons were already meeting with Barksdale's medical staff to arrange the handling of any crew remains that might be recovered and brought to the base.[19]

FEMA's Scott Wells arrived and located Whittle, who was swamped with trying to establish order. Whittle challenged him: "Who are you, and what the hell are you doing here?"

"I'm Scott Wells, and I don't know what the hell I'm doing here. But I'm here!" That broke the tension. Wells immediately began assessing the situation to see what support FEMA could provide. Within hours, FEMA brought in computers and phone lines to enable Whittle's team to begin tracking debris sighting reports.

MIT members set up the situation room. An airman brought in a set of sectional maps of East Texas and Louisiana and tacked them onto the wall. As debris reports came in, the locations were marked with color-coded pushpins on the map. Johnson Space Center and the Lufkin disaster field office were on the phone with the MIT almost constantly.

Whittle directed that every piece of debris be photographed in place and its GPS position recorded. The information would be entered into a computerized database to enable analysts to plot what debris was falling and where. This capability would be important in directing searchers to look for specific items or for crew remains. Whittle also required that everyone sign in upon entering Nose Dock 6, the aircraft hangar where the debris would be processed. This would ensure the integrity of the investigation's debris accounting system.

**Accident Plus Eleven Hours**

A series of aircraft delays pushed the scheduled mid-afternoon departure of my Rapid Response Team from Kennedy Space Center to eight o'clock that evening. The C-141 cargo plane finally arrived, and we loaded everything through the ramp at the back of the plane. The seventy-nine of us buckled ourselves into canvas sling seats lining both sides of the cargo hold. Sitting on rollers mounted down the middle of the plane were two pallets of pre-packed supplies, safety equipment, protective suits for handling hazardous chemicals, and tools that might be needed by the first wave of recovery personnel arriving at a shuttle accident scene.

I tried to collect my thoughts about what our team would be doing once we reached Barksdale. The situation was changing so rapidly that I still had no feel for the scope of the task awaiting us.

As the leader, I felt the need to say something to the team over the plane's PA system in this moment of sorrow and anxiety. I offered a few impromptu words that I hoped were comforting and motivational, but which I am sure were inadequate to convey the depths of the feelings everyone shared.

I was exhausted, numb, apprehensive—and flying off to lead a mission unlike any for which I had ever prepared.

I sat next to KSC security director Mark Borsi. We quietly discussed whether the accident might have been a terror attack. I was certain that it was not. Borsi agreed, saying that it would have been exceedingly difficult for a terrorist group to plant a bomb that could have survived launch and orbit and then be activated during the shuttle's landing. The technical challenges of doing that were just too overwhelming.

A few minutes later, the pilot invited me, as the ranking passenger on board, to take a more comfortable seat up front in the cockpit. I accepted his invitation, but later regretted it. I wish to God I had stayed down in the cold and noisy cargo hold with my people.

*A good leader should know better*, I chided myself.

I looked around and noticed that there was another empty seat. I asked that astronaut Jerry Ross join us in the cockpit. Jerry had been dragged through the depths of hell that day, losing his friends and then having the grim duty of telling their families that their loved ones had perished. Jerry would normally have flown back to Houston with the joyful families and the crew after a successful mission. Now, as the designated Flight Crew Operations member of the RRT, he had a very different task.

I could only imagine what he must have been feeling.

The C-141 taxied out to the runway and lifted off into the darkness. Eleven hours after the accident, the KSC team was finally underway.

The people sitting in the cargo hold were still trying to come to grips with what was happening. Most of them had never been in a military plane before, and they were unprepared for the noise and cold. No one knew what to expect once they landed—or even exactly where Barksdale was.

We landed at Barksdale Air Force Base at about 9:30 p.m. local time—now more than thirteen hours after the accident. Everyone seemed tired and confused. I overheard the pilot and copilot debating about where they should park the plane. Neither of them had ever been to Barksdale.

Colonel McGuirk, the deputy base commander, met our plane and offered his support. "We're going to stand up an ops center for you, with data lines and anything else you need."

McGuirk had one warning for us: "When you walk out toward the hangar, there's a red line painted on the tarmac and the taxiways. If you walk across that red line, be forewarned that there's a very high probability that you will be shot."

These guys meant business. The air force did not want anyone going near the B-52 bombers or their armament. For the civilians on the team, McGuirk's admonition escalated the tension in what had already been a stressful day.

We boarded air force buses to Shreveport's commercial airport, where we rented cars. We then drove to hastily arranged rooms at two nearby motels.

## Accident Plus Fifteen Hours

During the course of the afternoon, partial remains of some of *Columbia*'s crew members had been found along a fifteen-mile track from San Augustine County into Sabine County. The overall area in which the ship's debris fell in Sabine County was twenty-six miles long, completely traversing the county and crossing the Toledo Bend Reservoir that lay along the Louisiana border. Using the debris sightings, the weather radar images he captured earlier in the day, and the location of three findings of crew remains that day, Greg Cohrs from the US Forest Service drew a line through the apparent centerline of the plotted points. It followed a track from west-northwest to east-southeast—from 298 degrees to 118 degrees.[20] The Texas Forest Service generated a similar search line that evening for Nacogdoches County based on what their teams had tagged or collected in the field.[21]

At 11:30 Central Time, Cohrs recommended that Sabine County crews conduct grid searches the next day on either side of the centerline he had plotted, starting at the places where crew remains had been located.

Law enforcement officer Doug Hamilton hated calling off the searches for the night, knowing that there were still astronauts to be found. However, it was dark and cold, and everyone was exhausted. It was time to try to get a few hours of sleep. The incident command team agreed to convene before daybreak at Hemphill's VFW hall.

## Accident Plus Sixteen Hours

The Rapid Response Team leaders and I found our motel somewhere near Shreveport. I can't recall with certainty where we were. We tossed our bags into our rooms and went back to the common area to plan for our meeting with Dave Whittle the next morning. We wrote down a few notes and started drafting procedures for workers in the field to deal with hazardous materials.

By this point, people who had been running on adrenaline all day were burning out. At about one o'clock in the morning—seventeen hours after the accident—I told everyone to go to their rooms and get some sleep. We would reconvene in five hours.

Exhausted from their long and dreadful day, everyone was appalled at the conditions in their motel rooms. Mark Borsi's room was "beyond shabby." He felt that if it were shown on TV, people would think that the condition was being overstated for dramatic effect. The bedsprings were broken, and the blanket had holes in it. Borsi was glad that he and the team were carrying weapons, because "we were in a nasty part of whatever town that was."

I went back to my room and flipped on the television out of habit, and there it was again—*Columbia* breaking up in the sky. All the day's suppressed emotions finally poured to the surface. I had maintained a stoic facade all day for my team, as I was taught a leader should do. Now, in the privacy of my room, I lost it.

## Chapter 6

# ASSESSING THE SITUATION

I met with the discipline leads on the Rapid Response Team in our Shreveport motel lobby for coffee at six o'clock on Sunday, February 2, and then we drove to Barksdale to meet the other initial responders and begin our work—whatever that might be.

We hit a snag almost immediately. A guard denied us entrance at the Barksdale main gate because we lacked military identification. Jerry Ross, a retired air force colonel, showed his ID and requested that the guard talk to the appropriate officer, who quickly resolved the situation.

I found Dave Whittle at Nose Dock 6, introduced myself, and offered the services of the Rapid Response Team. I said, "Dave, I report to you." By grade level, I was the senior leader, but Whittle had the larger role as the chairman of the Mishap Investigation Team. My "paper" seniority was irrelevant. He appreciated my straightforward approach, and we got to work.

Dave chaired the first Mishap Investigation Team meeting at eight o'clock. My management team and I were there, as were Scott Wells from FEMA and Robert Benzon and Clint Crookshanks from the National Transportation Safety Board (NTSB). NASA's Dave King and Jim Wetherbee conferenced in from Lufkin. Representatives from Homeland Security, the FBI, the armed forces pathology team, and state emergency personnel from Texas and Louisiana also called in.

Ralph Roe, head of NASA's Orbiter Project Office, summarized the data received from *Columbia* just before communications were lost. Temperature sensors showed heat inside the left wing. A signal—possibly spurious—indicated that the left landing gear was down and locked. (The landing gear are not supposed to be extended until a few seconds before touchdown.) Then the wing sensors failed, and seconds later all communications were lost. These were the only indications of trouble. The NTSB representatives cautioned that everyone needed to keep a completely open mind during the investigation. They said that no one should form any opinions at this stage of the process, because it might cause people to discount evidence or miss potential areas for investigation.

Dave King established three overarching priorities for the upcoming search-and-recovery operations. First was to protect public safety by finding and collecting hazardous debris. Whittle's Mishap Investigation Team, supported by my Rapid Response Team, was responsible for this effort. Second was to locate and recover the remains of *Columbia*'s crew. Jim Wetherbee and the astronaut corps were NASA's leaders for this task. Finally, searchers needed to find and recover data recording devices and sensitive communications equipment. These memory devices—cameras, computers, and "black boxes"—might contain information salient to the accident investigation. Everyone in the field would need to be alerted to look out for these items.

The NTSB's Bob Benzon, whose experience included the investigation of the Pan Am 103 terrorist bombing over Lockerbie, Scotland, in 1988, estimated that no more than 10 percent of the shuttle would make it to the ground. He believed that generally only the densest components would survive reentry. There was no precedent for making a more accurate assessment. Nothing such as this had ever happened before.

A glance at the hundreds of pushpins already marking debris sightings on the wall-mounted map of Texas confused me. If Benzon's estimate was right, how could we be receiving so many reports of debris along a path more than 250 miles long? This didn't look like the kind of crash scene we were expecting.

Once the meeting adjourned, Whittle, the other managers, and I discussed what we could do to gather intelligence about materials found in the debris field. We decided to send out several members of the RRT to investigate at various locations—Dallas, Lufkin, Sabine County, NASA's weather balloon research facility at Palestine, and Louisiana.

Meanwhile, back at Kennedy Space Center, the Mishap Response Team was in action. Led by Denny Gagen, this team provided the logistics support for whatever we needed in the debris recovery effort. One of their first tasks was to send over the next wave of personnel from Kennedy to supplement our recovery forces. We needed them immediately.

That morning, our security director Mark Borsi worked with the air force to set up the temporary morgue in a hangar bay. The team immediately went to work outfitting it in a manner appropriate for the solemnity of its intended purpose. They procured American and Israeli flags, as well as refrigerated storage facilities. Borsi was deeply grateful for the air force's commitment to provide whatever NASA needed for our fallen astronauts.

—

In the piney woods of East Texas one hundred miles due south of Barksdale, Hemphill's population had nearly doubled overnight, as news media, officials, and volunteer searchers made their way to the town. News of the sightings of crew remains had spread through the area. People showed up at the Hemphill fire station, offering to help search for *Columbia*'s crew.

Dwight Riley was among those eager to be of assistance. A resident of Sabine County for all of his sixty-five years, he knew the area as well as anyone. State troopers at the fire station directed him to the

VFW hall, which had been pressed into service as the staging area for volunteers.

At the hall, Riley observed "hordes of people" in sneakers, lightweight pants, and other clothing that he knew would not be any match for the conditions in the field. He located the woman who was assembling search teams and volunteered his services. "I might be old," he told her, "but I'm up to the challenge."

As Belinda Gay, head of the VFW Women's Auxiliary, served breakfast to some of the volunteers, she listened to the stories about the crew remains recoveries the day before, and learned that the command center urgently needed more volunteers. She told her husband Roger that she felt compelled to join them. He encouraged her to do what she felt she needed to do. Belinda walked with the search teams for the next three days, and Roger stayed behind to run the food service operation.

Sixty miles to the west, leaders from myriad state and federal agencies had poured into the Lufkin Civic Center overnight and into the morning. The presence of so many type-A personalities and the significance of the event electrified the atmosphere in the building. Everyone wanted to be useful. Many felt compelled to be in charge of something.

Fortunately, experienced leaders knew chaos was part of the normal process in the hours following a catastrophe. It would take forty-eight to seventy-two hours to gather the appropriate situational awareness, sort out priorities, and begin taking control of the situation. Until then, things would be messy—and probably get messier.

The nonstop activity since the accident caused people's perception of time to become fluid and deceptive. Mark Stanford from the Texas Forest Service found that working in the windowless building was like being in a casino, without visual cues to indicate how much time had passed. At one point, Stanford told someone that he had been on duty for seventy-two hours straight. He was surprised to learn that it was still less than twenty-four hours after the accident.

NASA and FEMA debated initially over who was in charge of the recovery operation. Two men from FEMA asserted to NASA's Dave King that they were in charge, since President Bush's disaster declaration designated FEMA to handle the aftermath of the accident. King felt NASA had given him the responsibility. King called NASA headquarters. Shortly afterward, the White House Situation Room called King to ask what was going on. A few minutes later, King was told, "You're in charge of the search and recovery. FEMA's there to support you."

Almost immediately afterward, one of the FEMA men came to King and asked, "What can we do to help?" No egos were involved—it was just that the roles needed to be clarified and confirmed. Multiple leaders simply could not all have the final decision-making authority.

It took several days to iron out the specifics, but everyone ultimately agreed that FEMA was the lead agency to respond to the declared national disaster. FEMA funded the operation and coordinated all the federal, state, and local support. NASA was the lead agency for information and intelligence. NASA was also in charge of the technical aspects of the search-and-recovery operations and the accident investigation. FEMA tasked the US Environmental Protection Agency to work with the other agencies to ensure public safety by collecting, decontaminating, and transporting the debris. The FBI was the lead agency for recovering human remains. Finally, the Texas Forest Service was the lead state agency for Texas, providing planning and logistics support to the overall operation, and they were the primary interface with the local incident command teams.[1] Other federal, state, and local agencies would lend support as required in their areas of expertise.

Lufkin served as FEMA's operational disaster field office for all operations, including staging assets and deploying search teams to the field. Barksdale was NASA's investigative center and debris processing facility.[2]

Only twenty-four hours after the accident, dozens of agencies and hundreds of people were starting to make sense out of an unprecedented disaster that spanned huge swaths of several states. And the scope of the problem was still unknown.

—

By mid-morning, eighty searchers were on hand at Hemphill's VFW hall, including volunteers from the local community, Department of Public Safety troopers, and personnel from the nearby US Forest Service Sabine, Angelina, and Davy Crockett Ranger Districts, as well as representatives from the Sabine River Authority.[3] Search coordinator Greg Cohrs divided the people into four teams of twenty. He selected crew leads based on their leadership, "woods-worthiness," and navigational skills. He assigned the teams to grid-search two areas near where the first significant crew remains had been found. The crews would search along either side of the centerline Cohrs had drawn the previous evening, where it appeared most likely they would find additional remains. Groups One and Two walked northwest along both sides of the centerline from Beckcom Road. Groups Three and Four worked their way northwest from the location of the crew remains found on Farm Road 2024 near Housen Bayou.[4]

Grid searching is a basic skill for forestry workers—a methodical, disciplined way to search an area for still-smoldering pockets of embers that might rekindle into flame. Law enforcement also uses the technique to search crime scenes for evidence or to cover a large area when looking for a missing person.

The process involved walking in a line with searchers spaced out abreast at about an arm's length from one another. In practice, separation varied from five to twenty feet depending on the thickness of the local vegetation. As the line moved slowly forward, searchers would scan the ground and trees for anything of interest, and immediately flag and record its GPS position. If crew remains were found, law enforcement officers would call in the astronaut recovery team. Any shuttle debris would be left in place, because of potential hazardous chemical contamination, unless the item appeared to be an avionics box or something personal from the crew. The search line would continue to follow its prescribed path until it reached a road crossing, where the group's end-line exit points were flagged. The command center kept track of the searched areas on a topographic map.

Marsha Cooper of the US Forest Service was on one of the teams searching between Beckcom Road and Springhill Road, a few yards from where the first crew member was found the previous day. Cooper was wearing her new $300 fire boots.

Her team stepped off Beckcom Road, over some brush, and into the field. She immediately halted in her tracks. "There's something here," she called out. The line boss asked her to repeat what she'd said. "There's something here," she said, her voice breaking.

She did not want the nearby news media crews to hear that she was standing amid human remains.

Dense briar thickets, fences, streams, and other obstacles tormented the searchers as they made their way through the forests. The search lines had to go over or through those impediments—not around them—to ensure that nothing on the ground was missed. It was brutal work. Brambles and thorns shredded clothing and drew blood.

Volunteer Dwight Riley saw some people give up after a hundred yards, when their clothes proved woefully inadequate. One FBI agent, attired in a business suit and dress shoes, stepped in deep mud. He had to be extricated by his teammates, who then had to retrieve his shoes from the muck. DPS troopers' highway patrol uniforms afforded no protection from the environment.

Belinda Gay, who had been serving breakfast to the searchers, found walking the search line a strangely comforting—and somber—experience. There was little talking other than the occasional admonition from the leader to stay in line. No one knew what they were going to find, although they were keenly aware of the possibilities.

Debris of all sizes littered the area. Items from the crew compartment sobered the searchers. Pieces of mission patches—some scorched, others nearly pristine—lay in the fields. Golf balls and other items flown by the crew as presents for friends and family also turned up. Pieces of the crew's flight suits—boots, glove lock rings, and helmets—gave people pause.

In addition to the organized searches, local residents were requested to walk their property to look for anything that might have come from the shuttle. Hivie McCowan, an elderly widow, was deathly afraid to look around, because she feared she might encounter human remains. But she mustered the courage to walk through her pasture, and she found a large metal beam that clearly did not belong there.

The US Forest Service secured permission to resume helicopter flights over the area. Don Eddings of the Sabine Ranger District flew as a spotter in a contracted Bell 205 helicopter. His goal was to locate and record the position of possible crew remains and debris in inaccessible locations. As his helicopter flew between Bronson and Spring Hill Cemetery, something on the ground reflected the morning sun. He asked the pilot to land nearby, and Eddings found a large plastic envelope fastened with Velcro. Curious, he opened the envelope. It contained what appeared to be *Columbia*'s flight plan. He marked the location and called it in for pickup.

In Lufkin, astronaut Jim Wetherbee contemplated his responsibility for recovering *Columbia*'s crew. His assignment was clear: *Find the remains of the crew. Do whatever it takes. There are no rules.*

The potential scale of the search effort required seemed almost impossible to comprehend. Analysis of the tracking data during *Columbia*'s descent and breakup showed that debris rained down over a three-hundred-mile-long, fifty-mile-wide path stretching from Dallas to Fort Polk, Louisiana.

With locally led searches already underway in East Texas, Wetherbee established a military-style command and control system, led by astronauts, to solidify the search operations. John Grunsfeld was "ground boss" in Lufkin, charged with planning the ground searches each day, coordinating the search teams, and using feedback from the field to refine the overall strategy for subsequent days. He was also responsible for narrowing the search field based on analyses of recovered material.

Scott Horowitz, as the "air boss," managed all airborne searches for the crew.[5] Steve Bowen and Jim Reilly oversaw the water search operations in the lakes along the debris path.[6] Marsha Ivins coordinated the administrative and logistical support operations for the astronaut office team in Lufkin.[7]

Their leadership operations were completely separate from the activities associated with retrieving *Columbia*'s debris. Wetherbee and his team did not discuss their activities with the teams working on the debris recovery effort. Information did flow in the other direction, though, as locations of retrieved items from the crew module might help target people who were searching for the crew.

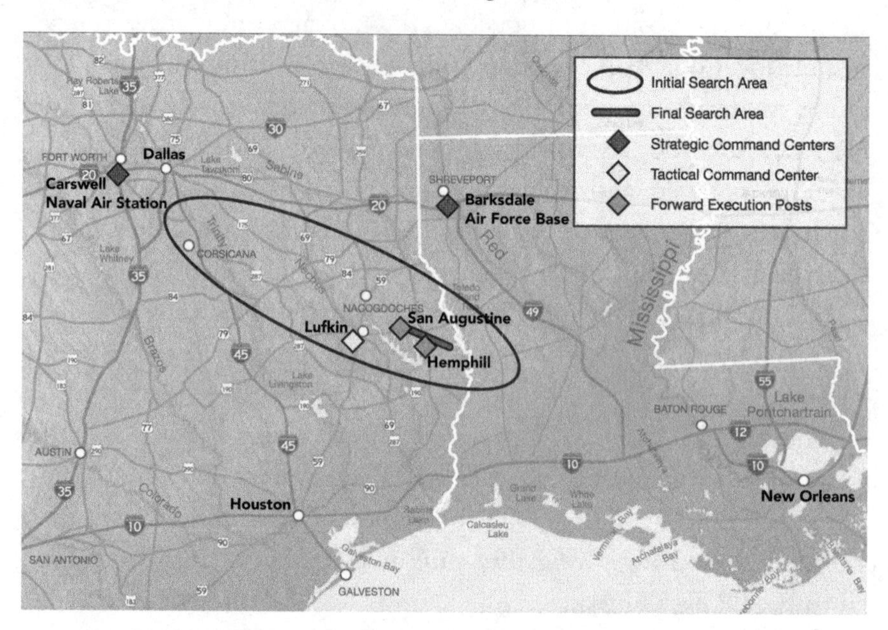


Map of East Texas showing the initial search area (ellipse) based on the first reported debris sightings. During the course of the following week, NASA gradually refined the search area for *Columbia*'s crew to a narrow corridor between San Augustine and the Toledo Bend Reservoir.

As the day progressed, it became clear that search teams in Sabine County needed more GPS equipment. Over the next several days, the command center sent people to every Walmart, sports store, and outdoor outfitter within a 150-mile radius to purchase all the available handheld GPS devices. The team also bought $35,000 worth of equipment and supplies from Hemphill's only office supply store.[8] Other command center staff rustled up first aid supplies from the local pharmacies and the Walmart in Jasper.

Leaders of NASA's command teams from Lufkin arrived at the Hemphill command center. They asked to see law enforcement officer Doug Hamilton from the US Forest Service, who had been photographing debris with his digital camera over the past two days. After reviewing his images with them in private, he drove them around to the locations of various items they wanted to examine firsthand.

Astronaut John Grunsfeld returned to the Hemphill command center to help plan searches for the crew. He mentioned to Olen Bean of the Texas Forest Service that his laptop lacked mapping software, so Bean offered to let him use a copy of a digital street atlas. Grunsfeld tried several times without success to get the program to load on his computer. Finally, Bean pointed out that Grunsfeld was hitting the "X" (close box) instead of "Yes" to complete the installation. Grunsfeld, a man with a PhD in physics—someone who had repaired and upgraded the Hubble Space Telescope on two separate missions—smacked his head and said, "Man! I'm so stupid!" It lightened the mood. But it also bore witness to the enormous pressure he was feeling.

Other astronauts arrived at the command center to help investigate sightings of possible human remains. Among them was Scott Kelly, Mark's identical twin brother.

Shortly after noon, the command center in Hemphill received a call that a crew member's body had been found deep in the woods near the Yellowpine Fire Tower, in an area known locally as Seven Canyons. The landowners searched their property that morning using all-terrain vehicles and found the remains near the boundary of Sabine National Forest.[9] Mark Kelly, Terry Lane, and two Texas game wardens went to the property to investigate. The landowners led them through the dense forest. It took nearly half an hour to reach the site using the ATVs. Kelly and his team saw the astronaut's body lying on a small mound in a clearing almost one mile from the closest roadway. The beauty of the surroundings belied the tragic nature of the situation. Kelly and Lane sat with the fallen astronaut for almost ninety minutes as they awaited the arrival of the rest of the recovery team.

Brother Fred Raney delivered a prayer, and the FBI documented the scene. While the recovery was in process, a call came in that another body had been located about five miles away, on the other side of Hemphill. This sighting was near Housen Hollow Lane, west of Farm Road 2024. After carefully bringing the crew member's remains out from the deep woods at Yellowpine, the recovery team headed off toward the Housen Bayou site.

Television satellite trucks massed along the street across from the Hemphill Volunteer Fire Department's fire hall. The command center was cordoned off, but incident commanders had difficulty entering and leaving the building undisturbed. Co-incident commanders Sheriff Maddox and Billy Ted Smith in particular were being closely watched. They had to resort to stealthier tactics—changing shirts or jackets, leaving through the back entrance of the fire hall, or using decoy vehicles—just to get their work done.

To be fair, the national newspapers and television outlets showed respect by not sensationalizing information that might be upsetting to the families or to NASA. Still, they were eager to tell the world—and the world wanted to know—what was going on in this high-profile search-and-recovery operation. They pushed the boundaries until they were told to back off. Reporters attempted to follow searchers out into the field. Police sometimes had to hold them back to keep them from interfering.

Much to the relief of the incident commanders, the local populace was circumspect about sightings of human remains on their property—calling the command center, not the press. To this day, the people of Sabine County will not discuss with outsiders what they saw of the remains of *Columbia*'s crew.

However, the media somehow got wind that a third crew member had been found in Housen Bayou. Texas DPS officers held most of the reporters back about a quarter mile from the entrance to the woods, but one reporter had arrived on the scene before the roadblock was established. He told Sheriff Maddox, waiting by the ambulance dispatched to recover the remains, that he had a constitutional right to follow the recovery team back into the woods. After arguing for several minutes, Maddox presented the man the option of leaving the scene voluntarily or being taken out in the back of a police car.

While Maddox and the reporter argued, the landowners took Kelly, Lane, and Brother Fred to the astronaut's location. The mud and briar thickets made the swampy bayou area difficult to navigate. Kelly went forward to be near his fallen colleague. After pausing in reflection for a moment, he motioned for Raney to join him. Raney once again read a few passages from Scripture and said a prayer. It was an arduous and sorrowful process to carry the astronaut's remains back out of the woods to the waiting ambulance.

—

Ed Mango, Dom Gorie, Jerry Ross, and several specialists from the Rapid Response Team left Barksdale Air Force Base for Lufkin at noon in their rental cars. Upon arriving in the Civic Center a few hours later, they were struck by the chaotic activity. Ross's and Gorie's blue astronaut flight suits immediately drew attention and inquiries from people who sought direction in how to handle the debris reports. Someone said the governor's office was insisting that NASA remove debris from all the school and hospital grounds as quickly as possible. Ross gave his car keys to one of the three men who came with him from Barksdale and asked them to help with the school cleanup.

Ross staked a claim to a work area in the midst of the chaos. He began to formulate a plan and establish the working relationships that would help the debris team accomplish its task. Then he started brainstorming the critical questions he would need to answer: *How do you conduct a methodical search? How do you photograph, document, and mark the location of each item consistently? How do you ensure that it's not contaminated before it's collected? What are the logistics for getting each item from the field to some holding area and then to wherever it is eventually going to end up?*

Ed Mango asked one of the Texas Forest Service workers if he could see a printout of some of the debris sightings. He noted an entry regarding a tire found at a farm near Chireno, Texas. He wondered how a rubber tire could possibly survive the heat of reentry.

Then they heard about an unusual sighting near Fort Polk, Louisiana. US Forest Service personnel, looking for debris after hearing loud sonic booms the previous day, had driven past several water-filled mudholes in the remote forest. At first glance, the holes were not particularly noteworthy. But then the searchers saw that mud was splattered forty-five feet high on the trunks of trees surrounding several of the holes.

The holes were impact craters.

The team agreed that they needed to investigate these unusual sightings. Dom Gorie had secured for the team's use several Army Reserve helicopters, which were now stationed at the Lufkin airport. Mango would take a helicopter eastward along the debris track from Lufkin, and Gorie would go west from Nacogdoches. Ross would investigate the craters in Louisiana.

—

At the two o'clock status update meeting, Bob Hesselmeyer from Johnson Space Center was announced as the leader for a Data/Records Working Group, tasked with analyzing telemetry and other data from *Columbia* to try to understand what caused the accident. There were already indications that thirty-two seconds of additional information from *Columbia* might be stored in Mission Control's computers.

Meanwhile, NASA had asked the public for help with the *Columbia* investigation. They requested that people contact NASA if they discovered debris or had film or video evidence of the accident.[10]

The EPA deployed hazmat teams to the field to help decontaminate and collect hazardous debris. EPA was also flying an aircraft with infrared sensors to detect the presence of hazardous chemicals.[11]

Six EPA-led search teams began searching for debris in schoolyards along the debris path, assisted by NASA engineers, the Texas Commission on Environmental Quality, and local volunteer fire departments. Fifteen teams would be available the next morning to concentrate on clearing school grounds. In Nacogdoches County, volunteer firefighter Jan Amen told a friend, "Never in my wildest dreams did I picture myself scouring the school yard for pieces of a spacecraft."[12]

—

Ed Mango was on his way to investigate the mysterious sighting of a tire in a field near Chireno, Texas. As he prepared to board his helicopter at the Lufkin airport, he saw another military helicopter land about a hundred yards away, met by a color guard and an astronaut. All activity ceased as the remains of a *Columbia* crew member were solemnly carried from the helicopter to a waiting ambulance.

Mango's Blackhawk helicopter crew included the two Army Reserve pilots, a technician from United Space Alliance, a Texas state trooper, and a forestry worker. Mango gave the pilot the tire's GPS coordinates and suggested they get to the site as quickly as possible. It was already late in the winter afternoon.

They flew low over the national forest on the way to the farm where the tire had been reported. They could see dozens of pieces of debris littering the ground along their route—so much that it was obviously not just trash strewn by careless people. At first, the team recorded the GPS coordinates of each sighting. After logging ten items, they realized there was no way they could keep records of every sighting and still make it to the tire before sunset, so they pressed onward.

They finally saw the tire in a clearing, and from the air it appeared to be relatively intact. The residents of the farm—a woman who appeared to be in her forties, two kids under ten years old, and an older man who appeared to be their grandfather—had come outside after hearing the helicopter circling overhead. The chopper landed, and the pilot walked toward the family to ask their permission to land on their property. His bubble helmet and green flight suit must have appeared frightening, because the children cowered and clung to their mother as he approached. Mango, dressed in less threatening civilian clothes, ran over and explained that they were federal representatives and would like her permission to investigate the debris on the property. She said, "Of course you have my permission!"

Mango explained that he would not be able to remove the tire from the property, and that the family should stay away from it in case it was contaminated by fuels. The mother asked, "Are you part of the army?" Mango said that he worked for NASA.

The grandfather appeared stunned. "You mean there's a NASA guy in our backyard?"

The mother then said something that stuck with Mango for the rest of his NASA career. It was a statement, not a question: "We are going to fly again, right?"

She also told him about some wreckage in the neighboring, unoccupied property. Mango's crew went to the spot and found a locker from *Columbia*'s crew module. Some of its contents were intact.

After taking photos, the team got back in the helicopter and returned to Lufkin, arriving just after sunset. Mango was sobered by the quantity and extent of debris he saw from the air during his brief helicopter trip. Only about a dozen NASA engineers and technicians were in Lufkin at the time to help with debris recovery. Mango called Barksdale that evening and asked our team to send as many people as possible to Lufkin the next day to help with the debris search.

While Mango was investigating the sightings in Texas, Jerry Ross flew from Lufkin to Fort Polk, Louisiana, aboard another Blackhawk helicopter. US Forest Service representatives met him and drove him by Humvee deep into the forest. When he arrived, he instantly recognized parts of the powerheads of *Columbia*'s main engines, still visible above the water filling in the craters. As the densest and heaviest components on the shuttle, the turbopumps had followed a ballistic trajectory at supersonic speeds from the time they broke away from the shuttle until they slammed into the muddy ground in the Louisiana forest. These components would later turn out to be the easternmost debris found from *Columbia*.

Ross and his guides examined two impact craters—one of which was located on the fourteenth fairway of the Fort Polk golf course—but were unable to visit a third site before darkness and rain set in. Ross flew back to Barksdale that night.

—

Jim Wetherbee knew that trying to find the remains of seven people in a ten thousand square mile accident zone was an intractable problem. The area was simply too big. Senior officials from Washington questioned why he had not immediately requested ten thousand National Guard troops to search for the crew. Wetherbee pushed back. It was impossible to bring in that many people immediately to search such a large area. There was no way to support them in the field even if they could deploy there. And ten thousand troops could not efficiently search that much land for crew remains and small pieces of equipment.

Wetherbee asked headquarters to be patient. If his team could find a way to narrow the search area, then he would request additional people as needed.

In the first day and a half, search teams spotted several hundred possible remains in the field. The vast majority later turned out to be from animals, but it would take several days for DNA analyses to confirm identification. Many of the suspected remains, though, were very likely from the crew. Wetherbee and Grunsfeld plotted their locations on a map. They decided to narrow the search efforts to an area sixty miles long and five miles wide, centered on the best-fit line of the suspected crew remains recovered on the first day. This would help them better target the search efforts of the National Guard units who were going to be deployed to the area starting the next day.

To help coordinate NASA's participation in the intensive search activities in the areas where crew remains were turning up in the searches already being conducted by the local incident command teams, Wetherbee established forward execution posts in San Augustine and Hemphill. He assigned astronaut Brent Jett as his forward coordinator in the field to brief the local team leaders each day.[13] The working relationship with the local search teams assured that NASA could request resources or assistance in searching an area, but direction of the ground searches resided with the local incident commanders.[14]

—

That evening, at the six o'clock status update to Dave Whittle and our team at Barksdale, Jerry Ross reported on his findings near Fort Polk. Bob Benzon of the NTSB agreed to conduct the follow-up investigation. We noted that the potential scope of the debris recovery operation was still growing. Possible debris sightings were being called in from British Columbia to central Florida.

The teleconference revealed the huge extent of the resources moving into position to assist with the search-and-recovery operations. Four helicopters would be available for searches the next day. Fifteen search teams would be deployed on Monday and thirty by Tuesday. Three hundred Texas Army National Guard troops were being deployed to the area for security and recovery of crew remains. The US Coast Guard deployed members of its Alabama Gulf Strike Team to Lufkin to assist with collection of potentially hazardous debris. Thirty-four additional personnel would be arriving at Barksdale from Kennedy Space Center late that night. Carswell Naval Air Station would be the temporary storage location for debris recovered between Corsicana and the Dallas/Fort Worth area. Other Texas temporary collection sites were in operation at Hemphill, Jasper, Nacogdoches, and Palestine.

The number one priority item for the debris search was the shuttle's Orbiter Experiment system (OEX) recorder. The OEX system monitored and recorded hundreds of channels of data during the shuttle's ascent and reentry. It was installed on *Columbia* to help characterize various physical loads on the ship, ranging from stresses and strains to temperatures and acoustics. It was the closest thing on the shuttle to the flight-data recorder found on all commercial airliners—and *Columbia* was the only orbiter instrumented that way. If it was intact, the OEX recorder might contain a data tape of *Columbia*'s status during its reentry and up to the time that it broke up. This would be crucial to help us determine the cause of the accident.

The meeting closed with somber news. Two ambulances were en route to Barksdale carrying remains from at least two crew members. They were expected to arrive at eight o'clock.

The Lufkin leadership team's teleconference with Texas Governor Rick Perry began at seven o'clock. Perry was upset that the state's school grounds had not been cleared of hazardous materials, since schools were scheduled to open the next morning. He demanded all school grounds be cleared by midnight.

The NASA and FEMA representatives briefly muted the line in order to talk among themselves. There was no way they could reliably clear all the schools in the debris corridor—the full extent of which was still not known—by midnight. The FEMA representative came back on the line and told Perry, "We need to put together a plan. We'll call you back in one hour."

The leaders quickly decided to assemble teams of three or four people to visit the schools in Lufkin, Nacogdoches, and Hemphill to determine what material had fallen there and how long it would take to clean it up.

The crews fanned out and reported back as best they could—cell phone coverage being spotty or nonexistent in wide areas of rural Texas in 2003. Five schools in Nacogdoches were reported cleared of debris.

There was no definitive news to give to the governor at the nine o'clock update call. By ten o'clock, there was enough data to inform Perry that it was impossible to open all the schools in the area the next day. The team suggested that the most prudent approach was to keep the schools closed, at least on Monday. Perry agreed. The next day, the team would make a concentrated effort to collect all the debris from the schools in the recovery zone.

—

As midnight approached on Sunday, the first full day after the *Columbia* accident, Marsha Cooper and her fellow Sabine County searchers were exhausted. They had endured a physically demanding and emotionally draining day unlike any in their experience. But they knew that the remains of most of *Columbia*'s crew members had yet to be located.

Cooper's team received instructions to investigate a call from a distraught woman who had found something on her property. She drove her US Forest Service colleagues Don Eddings and Felix Holmes to the woman's farmhouse, followed by several other cars of searchers. The woman told them she had stuck a tree limb in the ground to mark what she found. Cooper and her team walked far out into the pasture by flashlight until they saw the tree limb.

It was planted in the ground next to a pile of cow manure.

Cooper and her team burst out laughing. Their nerves were so raw from the long, horrible day that this situation seemed hilarious. "At that moment, we needed that pile of crap," Cooper recounted.

The searchers called it a day. Cooper went home. She had one thing left to do before she collapsed into bed.

Remembering her dreadful encounter with the crew remains at the start of her first search of the day, she took off her expensive new boots. She walked outside, placed them in the burn barrel, and set them on fire.

## Chapter 7

# SEARCHING FOR THE CREW

By the end of the day after the accident, operations had evolved into two parallel efforts, one for recovering the crew and the other for protecting the public from hazards and collecting the wreckage from the ship. Each task had its own distinct objectives and was led by different NASA personnel.

On the morning of Monday, February 3, the teams searching for the remains of *Columbia*'s crew gathered for breakfast and the safety and operational briefing at Hemphill's VFW hall, reviewing subjects including weather forecasts, search areas, and team assignments. Astronaut Brent Jett conveyed the plan from search headquarters in Lufkin that the search teams in San Augustine and Sabine should concentrate their efforts on the areas where it was most likely that crew remains could be found quickly.[1] He also highlighted the special shuttle components that NASA wanted the teams to look for. These included items such as the Orbiter Experiment system recorder, cameras, computers, and communications equipment.

Greg Cohrs now fielded six groups led by US Forest Service personnel, totaling about 175 searchers. The Texas National Guard deployed about 150 searchers divided into three search parties. Cohrs sent the Forest Service groups on either side of the area that was searched the previous day, between State Highway 83 and Bronson. The new Forest Service groups searched either side of the area covered the previous day near Beckcom Road. Cohrs deployed the National Guard teams in the Bronson area, headed northwest toward US 96.

Cohrs also had twenty-eight people, divided into fourteen teams, to respond to calls about debris sightings. Six FBI Evidence Response Teams of two people each responded to calls from the public or from the search teams regarding possible crew remains.

Military spotters would be aloft looking for broken branches in the forest treetops—evidence of things falling from the sky at a high speed.[2] Some of the Blackhawk helicopters carried canine search units to assist in locating crew remains.

Cold, light rain showers made footing treacherous in the field. Some searchers slipped and injured themselves. One woman fell and broke a hip; she had to be carefully extracted from the deep woods.

The ground searchers located two more of *Columbia*'s crew that day. Both were found in the Housen Hollow area between Farm to Market Road (FM) 2024 and FM 184.[3] The first call came in about three-thirty in the afternoon. The remains of the second astronaut were found nearby, while the response team was still in the area.[4]

As the response team went into the woods from the road to find the astronaut's remains, a white dog followed them, staying thirty or forty feet away. Sheriff Maddox assumed the dog belonged to the owner of the property. He worried about keeping the dog away from the crew members' remains, but this turned out not to be an issue. The dog stopped and lay down near Maddox while Brother Fred Raney read his words beside the fallen astronaut. To Maddox's surprise, the dog covered its head with its paws while Brother Fred led the prayer service. At the end of the service, the dog led the team out to the road. Then the dog went back into the woods and was not seen again. It did not belong to the property owner. No one knew whose it was or where it came from.[5]

Five of *Columbia*'s crew had now been located and recovered. Hopes ran high that the other two would soon be found.

Incident commanders in Hemphill experienced challenges in communicating with their people in the field. Cell phone coverage was spotty or nonexistent in much of the area. Billy Ted Smith asked his logistics chief, Mark Allen, to contact the wireless phone companies about installing temporary cellular service towers in Sabine County. Verizon agreed to provide the recovery effort with trucks carrying portable emergency cell phone towers, as well as a box of cell phones. These would be provided at no cost to the operation, with one request—that when Smith next briefed CNN, he would publicly thank Verizon for the donation.[6] Smith's public acknowledgment encouraged other wireless phone services to donate additional portable towers and cell phones.[7]

Now three days into the recovery period, the Hemphill command center was running at full throttle. The US Forest Service personnel and local law enforcement officers used their training in the incident command system to create clear lines of authority and ensure that important issues were being addressed. The leadership team built upon each member's training and expertise. Those who had never been through a major crisis quickly came up to speed.

At the same time, some of the command team leaders were ready to pass the responsibility for managing the recovery on to better-resourced organizations. Rumors circulated that the Department of Defense would take over the search operations.

When DOD representatives arrived in Hemphill on the third day, the command team greeted them and expressed their relief at being able to turn over the reins to the military. Much to their surprise, they learned that the officers were there only to inspect the operation. Everything appeared to be working well, so the military had no reason to take charge.

Inquiries about NASA, FEMA, or the US Forest Service assuming control were similarly dismissed. None of the forty-six federal, state, and local agencies on the scene saw any need to step in and take over an operation that was obviously working well.[8]

The Hemphill team was a victim of its own success at this point of the search.

—

Jim Wetherbee and his crew recovery leadership team in Lufkin established a regular operational routine. Reports from the field came in by phone or runner several times a day. Using supplemental information from databases and maps as they were updated, leaders in Lufkin had a good feel for the situation on the ground.

Every evening, Wetherbee's team planned their priorities for the next day's search and where to target the search teams based on the information available. Then they turned the plan into actionable tasks. Senior search leaders were briefed at night and the field leaders the next morning before executing the search plan. They called back with the results at the end of the day, starting the cycle over again. A feedback loop from field personnel to Wetherbee's team served as a management check. *Are you getting the support you need? What's working and what isn't?*

Despite all the organizational expertise being employed, Wetherbee said it felt like he had been making decisions every fifteen seconds for twenty-one hours straight for each of the first three days. He was getting about three hours of sleep each night—and it was not restful sleep. There was no time to let his emotions surface when he was on duty. But in his room late at night, he was alone with his thoughts—which constantly turned to *Columbia*'s crew and what had happened to them.

The intensity of the experience was already taking its toll on some people. Mark Kelly had been through the traumatic experience of recovering the remains of several of his colleagues, three of whom were his classmates in Astronaut Group 16. Kelly decided he had performed his duty. It was time to return home to be with his family and let someone else take over. Astronaut Nancy Currie was on hand the next day in Hemphill for crew recoveries.[9]

Wetherbee consulted with the FBI on how to help the astronaut searchers deal with the trauma of finding the remains of their friends. At the FBI's recommendation, he instituted a three-day limit for astronauts working in the field on crew recovery. He forbade his ops managers to participate in search operations, as they were already under enough stress. He also made it mandatory that astronauts returning from the field to Houston speak with NASA's resident psychologists.

Wetherbee developed a briefing to prepare arriving astronauts for the psychological and emotional agony of finding remains of their colleagues in the field. The primary responsibility of each astronaut was to recover the remains of *Columbia*'s crew with dignity, honor, and reverence. It was also imperative that the remains be taken to the temporary morgue as quickly as possible so that forensic evidence was not lost. Astronaut Dom Gorie would escort the remains to Barksdale.

Wetherbee instructed the astronauts to be guided by an "Eight, Eight, Eight Rule" in recoveries: "Eight days from now, eight months from now, and eight years from now, we must be able to live with the consequences of the decisions that we will make in the field. Every decision must be based on our highest judgment using our greatest professionalism and human values."[10] He was determined that things be handled with appropriate dignity and respect for *Columbia*'s crew. He knew the astronauts would do this without fail.

—

Hemphill's handful of small restaurants could accommodate only a limited number of people. As the staging area for the search teams in the morning, and the place where they returned after a day in the woods, Hemphill's VFW hall was about the only place searchers could go for a meal. And they came in droves.

Since so many people needed to be fed at the VFW hall, Belinda Gay realized that she was needed there more urgently than out searching in the field. She had extensive food service management experience from running her own catering business and "Fat Fred's" in town with her husband Roger. The incident command center officially deputized her on February 4 to run the food unit under Mark Allen's logistics team.[11]

People were already bringing in food and offering to help serve it at the hall, but there were overlaps and gaps. Belinda organized the volunteers, made hundreds of calls over the next several days, and informed the local radio station about the kinds of food needed each day.

Tuesday morning's good news was that all of the debris in or near the schools throughout East Texas had been cleared, and all schools could open. The bad news was that the search teams had been using school buses to get to and from the search sites, and now the school kids needed the buses. The Hemphill command center staff called the school districts in the neighboring communities to borrow additional buses. After taking the children to school, the buses were pressed into service to take search teams out to the field, returning to pick them up after kids went home for the day.

Hundreds of searchers had arrived in Hemphill over the past several days, and many were completely unprepared for the conditions. Personnel from Texas driver license offices were sent to help with the search, but they were not told what to expect. Many of these people had never been in the woods before. Some arrived without even a change of clothes. Texas state troopers were told to walk the woods in their regular duty uniforms, which were more suited to criminal investigations and highway patrol than to wildland operations. "Plantation" forests—cleared of older trees a few years before—were now young pine stands or dense brush thickets with briars that tore clothing to shreds. Sheriff Maddox remarked that some people had so many cuts and scratches that they looked like they had been through a meat grinder.

Ground search groups braved Tuesday's cold and rain to search the area from Farm Road 2971 south of Hemphill to US 96 near Bronson.[12] Floridians from Kennedy Space Center were not accustomed to the cold. Safety manager Gerry Schumann told his wife that it was like "hell on earth," between the cold, sleet, rain, and the briars. Before the week was out, thorns destroyed the three pairs of jeans he had brought with him. He passed the word back to KSC that people needed to be better prepared for what they would be facing.

Some of the searchers cut their way through the undergrowth with machetes. After a National Guardsman accidentally severed the end of his finger, machetes were forbidden on search teams. Experienced woodsmen preferred using "walking sticks." These large wooden poles could be used for support in slippery terrain, and they could push briar patches to the ground so that people could walk over or around them.

Diarrhea took its toll on dozens of people on the search teams. Days of sharing food and makeshift sanitary facilities contributed to its spread. Muscle cramps, dehydration, and exhaustion threatened to overwhelm some searchers.

The local volunteer firemen were on average much older than many of the people arriving from outside the area. However, they knew the deep woods intimately, and they were much better prepared for the conditions. Growing up in the area taught them many tricks for dealing with the undergrowth. Marsha Cooper wrapped one of her duty shirts in duct tape to help it stand up to the punishment.

Forestry technician Jamie Sowell from the US Forest Service led a team consisting of several other Forest Service personnel, a NASA representative, and as many as fifty volunteers from the local communities. He appreciated the dedication of the searchers, but he remarked that maintaining a straight search line with so many inexperienced people was often like herding cats.

Sabine County's volunteer searchers performed admirably in maintaining their search lines. They resisted the temptation to extend out the ends of their lines, which might have caused them to miss something in their assigned areas. When one team found crew remains only forty feet outside the area searched the day before, the searchers were naturally impatient to cover more ground. But Greg Cohrs and the team leaders reminded them that it was imperative that they maintain a disciplined, methodical search pattern.[13]

Sowell and his fellow team leaders actively monitored their teams to identify volunteers who might not be up to the task physically or emotionally. The leaders occasionally suggested that some individuals might be better able to help out in some way other than walking the woods. He ensured that people knew that there was no shame in admitting they were not suited to woodland grid searching.

One out-of-towner, dressed in a suit and driving a Mercedes, drove up to the VFW hall and offered his services. He changed into work clothes and gladly helped sweep floors, clean bathrooms, and take out the trash.

Townspeople were still adjusting to the worldwide attention their quiet community was receiving. Residents had a hard time believing that so many astronauts were in their small town. To ease the tension, the astronauts tried their best to be just another part of the family. When they were not involved in the recovery efforts, they went out of their way to talk with schoolchildren and the volunteers.

But the astronauts naturally attracted attention from the media. It was impossible for them to travel from the command center to the VFW hall for a meal without having microphones shoved in their faces. Kennedy's Pat Adkins, who was working in the Hemphill debris collection center, devised an order and delivery service from the VFW hall to enable the astronauts to eat in peace in the command center.

Housing everyone was becoming a logistical nightmare. Between people participating in a fishing tournament already underway on the nearby Toledo Bend Reservoir and all the searchers pouring into the area, no motel rooms were available within an hour's drive of town. Even the town's wastewater treatment facility struggled to keep up with triple the usual town population. Hundreds of National Guard troops were bivouacked in the Hemphill high school gymnasium, trying as best they could to make themselves comfortable and sleep on wall-to-wall cots.

Gerry Schumann was one of the few NASA personnel who was able to find a motel room in Sabine County. However, he slept in his car several nights during the first two weeks of the recovery period rather than make a late-night drive back to Hemphill from Lufkin.

Once again, county citizens came to the rescue. Owners of lake-front vacation houses in the area phoned the VFW hall to offer their properties for searchers' use. They refused to take any payment, saying that it was the least they could do to help with the recovery. Other people offered searchers free room and board in their homes and to do their laundry.

Simple acts of kindness and words of encouragement comforted the NASA workers who were still grieving over their fallen comrades and the loss of *Columbia*, and who felt so far from home. Local resident Pat Oden wrote to a friend, "I was making a video of one of the news conferences and started talking to a young NASA man. I asked if I could video him and he said to me, 'I am not anybody important.' I replied, 'To us you are very important.' As I was taking his photo, I asked him his name and was he from NASA Houston. He replied, 'No, ma'am, I am from Kennedy Space Center in Florida.' With that remark, I quit taking photos and walked over to him and hugged him and told him I knew how hard this must be on him. He started crying, as I did. Very emotional. Makes me feel good that I retired to this small community."[14]

—

The first of many memorial services for the *Columbia* crew was held at Johnson Space Center at noon on February 4. It was a private ceremony for family members, friends of the crew, invited guests, and NASA employees and contractors. President and Mrs. Bush attended, as did Sean O'Keefe.[15] At the ceremony, the president remarked that although NASA was being tested at this time, "America's space program will go on."[16]

In the skies above Russia, about an hour before the ceremony started, an unmanned Progress spacecraft docked with the International Space Station, bringing one ton of food and supplies to the crewmen—two Americans and one Russian. They could now survive at least until July without a space shuttle resupply mission. That was critical, because no one knew when a space shuttle might visit the ISS again.[17]

—

Admiral Gehman and other members of the *Columbia* Accident Investigation Board toured the three areas in the debris field and the collection area in the hangar at Nacogdoches airport. Gehman insisted the board see the debris in the field, to make the accident more personal to his investigation team and prevent it from becoming "an abstract event." He said that the board had two main responsibilities. The first was to future astronauts, who needed to feel that everything possible has been done to make it safe for them to fly. The second responsibility was to the three astronauts aboard the ISS, who needed the shuttle to fly again as soon as possible.[18]

Jan Amen from the Texas Forest Service was pressed into service as a photographer for NASA during Gehman's tour. Many of the photographs she took that day and over the next several months became part of the official record of the recovery operations. As a resident of a small town in rural Texas, she never dreamed she would be exposed to an event at the center of the world's attention. She wrote an email that evening describing her first day working with NASA as "one of the most incredible experiences of my life."[19]

—

Jim Wetherbee called upon Dr. Jim Bagian, a former astronaut who had helped with the *Challenger* crew recovery, for consultation on crew recovery procedures. Among Wetherbee's concerns was keeping any news about the status and condition of crew remains out of the press. Bagian helped him by developing a two-letter random code array to match the recovered remains to the appropriate crew member. Wetherbee was the only person in Lufkin with the full key to that code. This code enabled discussions about operations to be conducted over nonsecure channels without revealing sensitive information about which astronauts had been recovered and where.[20]

By February 4, it appeared that at least some portion of each *Columbia* crew member's remains had been recovered. Dr. Philip Stepaniak and his medical team in the temporary morgue at Barksdale received, studied, and prepared the remains of *Columbia*'s astronauts that had been gathered so far.

The medical team and the recovery leaders in Lufkin debated about the appropriate time to transfer the crew's remains to the Armed Forces Institute of Pathology at Dover Air Force Base. On one hand, consideration for the grieving family members argued in favor of waiting to move all seven transfer cases at one time. On the other hand, it was imperative to get the remains recovered so far to Dover as quickly as possible, so more thorough forensic analyses could be performed to determine the causes of death. And until analyses of dental records and DNA could be completed, it was unclear just how much of each crew member's remains had been recovered.

Wetherbee and Dr. Stepaniak jointly decided to send all seven of the caskets together on a flight to Dover the next day. They made the decision to ease the suffering of the families, the NASA family, and the public.[21]

At the end of the day's operations, the search team leaders debriefed with some of the command team members in Lufkin. They provided feedback on how search techniques were working and whether an area was "cold"—without any significant material being found. Armed with this feedback and the GPS positions of the material and remains being recovered, the command team planned the next day's searches.

—

On Wednesday, February 5, at the eight o'clock morning phone call, FEMA's Scott Wells reported that the Texas Army National Guard, coast guard dive teams, and six FBI canine teams were engaged in the search efforts. They were concentrating on a "hot spot"—a ten-mile by two-mile area—near Bronson and Hemphill. Wells noted that all federal agencies on the scene were at peak manning. Now the teams just had to "do the work."

The search teams continued to grow. Now five days after the accident, there were eight hundred National Guard troops on duty in the search corridor, most of them working on crew remains search and recovery. The Department of Public Safety had 353 personnel in the field, and there were 140 US Forest Service workers on site. The Texas Commission on Environmental Quality had twenty-three teams deployed, and EPA had 370 people deployed. Sixty EPA teams had collected more than 1,100 bags of debris.[22] The Texas Forest Service had 140 employees scattered throughout East Texas and on duty at Stephen F. Austin State University to provide support to the recovery.[23] Louisiana had 174 searchers in the field.[24]

In Sabine County, the same search groups who had worked the previous day now combed the area stretching from south of Hemphill to US 96 near Bronson.

Hemphill's Dwight Riley was participating as a volunteer searcher. He was becoming obsessed with the search in a very personal way, curious about the shuttle and its astronauts and how everything had come to rest in his county. Every find was seductive, something that made him empathize with the crew and the shuttle.

One day Riley found a "Lift-the-Dot" type of fastener on the ground. It made him wonder: *Where did that come from? Was that from a harness? Was it off a uniform? Was it used to fasten something to a wall?* On another day's search, he found a piece of curved glass, "smoky" on one side, which appeared to have an inspection seal on it. Again, he wondered where on the shuttle it had come from and why it was lying here so peacefully in front of him. It was difficult to fathom that each little piece had been in space, had gone through the breakup of *Columbia* forty miles up, flown across Texas, and then fallen to Earth in his county. Even though his "old bones were dog tired" every evening, and he was happy to have a hot shower at the end of a grueling day, each morning he was eager to don his Carhartt overalls and get back out in the rain to help with the search.

Mike Alexander, another local volunteer, felt the same way. Painful arthritis made him hesitant to search the first few days, and the going was tough. However, he found the experience so rewarding and compelling that he told his wife that he just could not stay home.

The US Forest Service team leaders appreciated the local knowledge, determination, and tenacity of these older volunteers. Jamie Sowell described Riley as "tough as a boot" and "a godsend"—someone he could depend on to help the newcomers.

In the fire hall, co-incident commander Billy Ted Smith felt that he was on borrowed time. As an employee of ExxonMobil in Beaumont, Texas, he had received permission to take off several days to work the incident command for the shuttle recovery. Now, he thought he needed to turn over the responsibility for the operation to someone else so that he could return to his day job. A congressman visiting the site told Smith that he was too valuable to the operation to leave. A few phone calls garnered Smith permission to stay in Hemphill as long as he was needed. He ended up being on duty in the command center for twenty-seven days.

With almost no advance notice and little time to prepare, Greg Cohrs and Marcus Beard were asked to step in front of the cameras and brief the national media at five o'clock. Reporters asked Cohrs several times about crew remains, but NASA had instructed him not to discuss anything about the crew. Cohrs and the incident commanders knew that the bodies of two of the crew members had not yet been recovered, seemingly contradicting NASA's press release that seven flag-draped caskets containing the remains of *Columbia*'s crew members were arriving at Dover that afternoon.[25] The NASA release was essentially factual—they had recovered partial remains of all seven crew members—but the bodies of two of them had not been located. Cohrs suspected that the media felt things did not quite add up, because they were aware of the continuing activity around the command post and the county.

After the news conference, four Native American fire crews arrived in town. Cohrs planned to deploy these additional eighty searchers the next day. Wildland fire crews like these, contracted by the US Forest Service, would become the backbone of the recovery effort within two weeks.

—

Thursday, February 6, was a miserable day in Sabine County. Temperatures ranged from 40°F to 44°F, with sleet and about an inch of rain. Swollen creeks, some chest deep, were difficult for searchers to cross. Sixty-four agencies were at work in the county, with about 850 people involved in search and support. Searchers received yellow rain slickers to wear over their heavier clothes. Briars and nearly impenetrable thickets shredded the thin material within minutes.

Greg Cohrs was concerned about the safety of the searchers, the risk of hypothermia or exposure, and the dangers of crossing rain-swollen creeks. But he had to find the remaining crew. Time was running out. The longer the bodies remained in the field, the greater the likelihood of forensic evidence being lost.

In the late morning, the incident command team decided to recall the teams led by the Forest Service. The searchers felt angry and betrayed to be pulled in from the field. The National Guard, better prepared for the conditions, continued searching.[26]

Roger and Belinda Gay had been sleeping at the VFW hall every night since the shuttle accident. They rose at three o'clock every morning to begin cooking eggs for the searchers. Today, the Gays and their helpers cracked and cooked more than 2,500 eggs.

Volunteers from the community continued to donate food. People brought in whatever they could. An elderly gentleman whose wife was bedridden prepared a plate of fried chicken and brought it to the VFW. When he saw the throngs of searchers at the hall, he said, "This ain't gonna be enough to feed anybody." Belinda Gay told him, "It's the little things that make the big picture happen. Had you not brought that chicken, those three men over there might not have gotten to eat today."

One woman of modest means brought in a cooked chicken, and she apologized profusely that it was all she could spare. She appeared impoverished, so the searchers took up a donation and purchased groceries for her to ensure that she had enough to eat.

Someone delivered a sheet cake decorated as an American flag with the names of the STS-107 crew. No one would serve it. Eventually it was covered with plastic wrap and put on display.

Pat Smith, an employee at the Shelby Savings Bank, appealed to her manager to help feed the searchers. They purchased several cases of fried chicken from the grocery store. As she and her manager drove up to the VFW hall to deliver their donation, they were astounded at the line of cars ahead of them and "little old ladies getting out with their peas, their dumplings, their corn bread, and their cakes. They'd just deliver it and drive on out." Smith broke into tears when she saw firsthand the outpouring of support in the community.

Third grade teacher Sunny Whittington felt guilty about not being able to help out in some significant way after hearing her students tell about their parents' volunteering. As she drove to work, she stopped at the grocery store and purchased supplies for making sandwiches. Her teaching partner saw her unloading her truck at the school and asked what she was up to. "We're going to make sandwiches!" she said. "School isn't just about reading and writing and math. It's about life." Her colleague immediately offered to have her class help, and then called her husband to purchase more food. The children in the two classes made more than five hundred sandwiches that day.

While the searchers appreciated the kids' offerings, what touched them most was that a child's handwritten note of encouragement and support accompanied each sandwich. Many of the volunteers and National Guardsmen wrote letters back to the school to thank the kids. Even today, searchers weep when recalling how deeply moved they were by what the students of the elementary school did for them.

Despite the community's seemingly endless capacity for good works, it was quickly becoming apparent the VFW could not continue to rely solely on the generosity of the local populace. The perishable food being brought in overwhelmed the storage capacity of the VFW's refrigerators and freezers. Health issues from unsafe food were a concern.

Sheriff Maddox called his friend Jerry Powell at Tyson Foods to ask for a loaner refrigeration unit. Powell said, "It's on its way. Do you need any food?" Maddox replied that they would gratefully accept anything Tyson could offer. Powell sent a freezer truck full of frozen chickens along with the refrigerators. A group from Gregg County brought steam tables to the VFW hall and began preparing hot food—a service they performed for the following week.[27]

The VFW hall had three serving lines, feeding more than one thousand people three times per day. There was so much food and so many searchers on hand that the volunteers had to feed people in shifts.

Hivie McCowan—the widow who had been afraid to look for debris in her pasture several days earlier—volunteered as a food server at the VFW hall. "To let them know what I was serving, I kept saying, 'Sweet tea? Sweet tea?' I'd said it so long that after a while, they just started calling me 'Sweet Tea.' It was awesome!"

And much to the relief of the early morning food service volunteers, the McDonalds in Jasper announced they would start sending prepared breakfast meals.

—

Astronauts Jim Wetherbee, Jerry Ross, John Grunsfeld, and Scott Horowitz tried to refine the search area for the crew's remains and the critical components inside the crew module. They were attempting to reconcile conflicting data, analyses, and speculation about how the crew module might have come apart and how winds aloft might have carried the remains and debris in various directions. It was confounding their attempt to narrow the search path.

Ultimately, Wetherbee and Ross told their teams to rely on the "ground truth," analyzing only the data for the items that had made it to the ground. A glove here, a helmet there—these were the physical evidence of how *Columbia* actually fell to earth. Their analyses, aided by geospatial staff at Stephen F. Austin State University, further narrowed the search path that day. They subsequently helped to target searches to precise areas where there was a high likelihood of locating remains of the crew.[28]

That evening, Brent Jett met with Cohrs and informed him that NASA wanted to realign the search centerline from the original one that Cohrs had drawn using the satellite imagery and the location of crew remains. Grunsfeld's analysis of GPS locations in the debris field enabled NASA to further narrow the search corridor for *Columbia*'s crew to an area one mile wide by twenty-five miles long. NASA thought the actual debris trajectory was four degrees different in azimuth from Cohrs's original projected line. Cohrs narrowed and realigned the next day's search area in Sabine County to correspond with NASA's analysis.

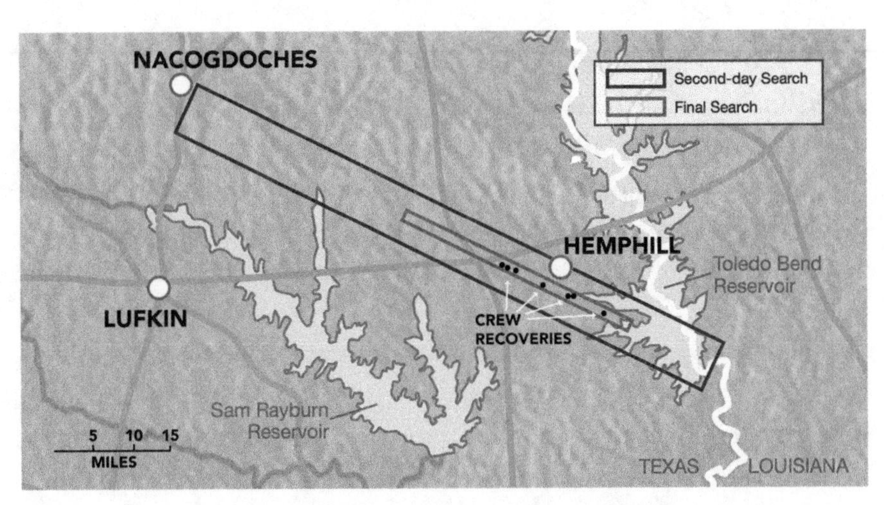


The remains of *Columbia*'s crew members were eventually found in Sabine County within the narrow search area defined by Jim Wetherbee and John Grunsfeld.

As February 7 dawned, Texas had 430 state employees active in the recovery. There were eight hundred National Guard troops on search duty. Louisiana had more than 175 state employees on search duty.[29]

Johnson Space Center encouraged its staff to volunteer to help in the search effort—but they would have to take personal leave to participate. Many JSC flight controllers, engineers, and other government and contractor personnel did take leave to help with the search, out of a sense of loyalty to their fallen friends and to America's space program. Six days after the accident, several JSC employees were on hand in Hemphill's VFW hall.

Greg Cohrs had about 650 people to put in the field in Sabine County this cold, showery morning, including six teams led by Forest Service staff, 230 National Guardsmen, fifteen mounted searchers from the Department of Criminal Justice, the eighty people on the Native American fire crews, thirteen dispatch teams, and six canine search teams. As the morning progressed, cold rain lessened in intensity and sky conditions improved, allowing aerial searches to begin again.

The incident command team discussed an offer of a thousand additional volunteer searchers for the weekend. The consensus was that the Hemphill command team's capacity was barely able to logistically support the feeding, transportation, and leadership span of control of the current number of searchers, let alone an additional influx of people equal in size to the town's entire population.[30]

The nonstop activity in the command center was taking its toll on Greg Cohrs. He likened the environment to being in a piranha tank, with people constantly grabbing him from every direction, sixteen hours a day. Despite the food being brought to the station by volunteers, Cohrs had no time to eat. He lost fifteen pounds in the week after *Columbia* disintegrated, and he was exhausted.

At eight-thirty that evening, as Cohrs was preparing instructions for the next day's search, Terry Lane from the FBI told him he was needed at the Lufkin center immediately. Cohrs jokingly asked if he was under arrest. Lane said no, but the leaders in Lufkin critically needed to hear from him personally regarding search-and-recovery operations. Cohrs hesitated, but Lane insisted. They departed for Lufkin thirty minutes later.

Having been confined to Hemphill's tiny operations center for a week, Cohrs was astounded at the huge scope and frenetic pace of the operations in Lufkin. He joked to Lane, "We could have been finished with the search by now if we had all these people!" At the command meeting, Cohrs was asked to give a briefing on his search operations. He laid out a taped-together map on the table and showed them the area already searched as well as plans for upcoming searches. He explained that although the teams were finding and identifying debris, they were not collecting any at the moment, because they were entirely focused on finding the bodies of the remaining two crew members.

A hush fell over the meeting. Someone said, "What do you mean, we're still searching for crew members?"

Cohrs replied, "We've only recovered five. We're still searching for two more."

Because of the secrecy wrapped around the crew recovery operations, the people in the command center who were involved only with the debris collection efforts were astonished to hear that two crew members were still missing.

Cohrs was unsure what the leadership team thought about his search plans. He feared the leaders would reassign his resources to San Augustine County, even though all the crew-related finds had been in Sabine County.

But the meeting concluded with the direction to redeploy San Augustine's resources to concentrate search efforts in Sabine County over the weekend. Cohrs would have a huge number of new searchers to coordinate the following morning. If there was one piece of good news, it was that the people coming from San Augustine already had several days of training and experience in the field.

Lane drove Cohrs home after the meeting. At about one in the morning, he collapsed into bed.

—

On Saturday, February 8, NASA's teams at Barksdale, Lufkin, and Carswell Naval Air Station briefly stood down to attend memorial services for *Columbia*'s crew. Meanwhile, a memorial service was held in Lufkin's First Baptist Church. NASA Administrator Sean O'Keefe spoke of NASA's gratitude to the citizens of Texas for their support in the recovery operations. Texas Governor Rick Perry also spoke, and astronaut Jeff Ashby delivered a tribute to his fallen comrades.[31]

—

In Hemphill, when Greg Cohrs and his colleague Paul Dufour drove to the VFW hall to deliver the morning briefing, they were shocked to see a line of state patrol cars stretching more than a quarter mile down the road. No parking was available within a ten-minute walk of the building. So many people were gathered outside that he had difficulty getting to the door. Cohrs found the inside even more crowded, "as if all of the people were stacked into the room standing up." As he made his way through the crowd, Cohrs ran into his son Adam, who had taken the weekend off from college in Beaumont to join the search.

In a fog of disbelief, Cohrs delivered the operational part of the briefing while he was still trying to figure out how to assign so many new resources—1,500 today versus 650 yesterday—and get them out to the field with the very limited transportation available.

He deployed the new teams to cover areas along the centerline that had not been searched before. But the searches were frustratingly inefficient. It was nearly impossible to maintain a disciplined search line with teams of fifty to eighty people. In addition, many searchers were not physically up to the task, and team leaders had to take them out of the grid before the end of the day. As important as it was to locate *Columbia*'s crew, it was not worth risking more lives to recover those who had already perished.

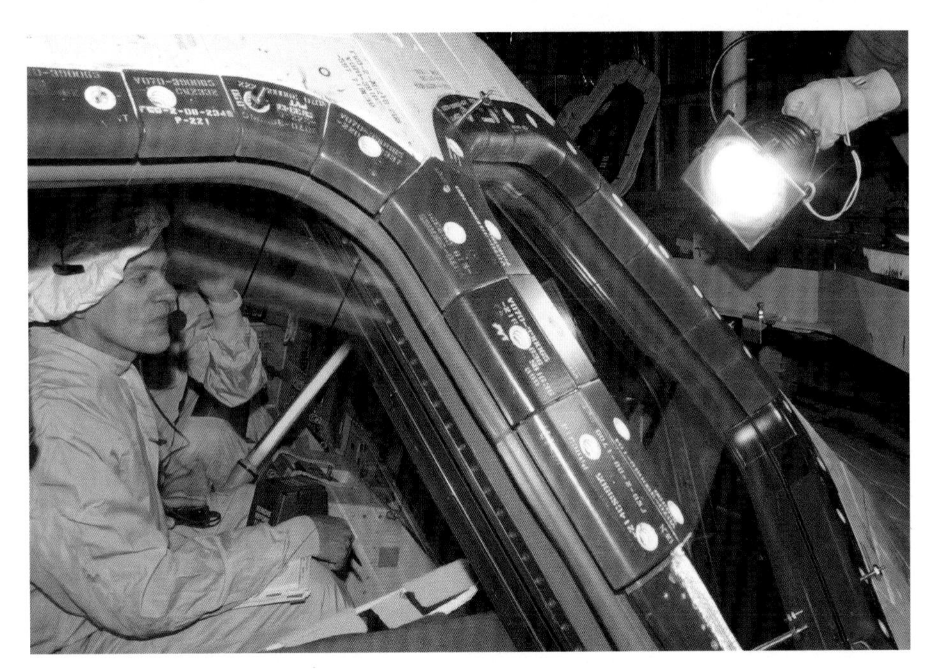

STS-107 pilot Willie McCool in *Columbia*'s cockpit during the crew equipment interface test, June 2002. Note the individually numbered black and white silica heat shield tiles. *(NASA photo)*

Test project engineer Mike Ciannilli with *Columbia* as it was being rolled from the Orbiter Processing Facility to the Vehicle Assembly Building. Ciannilli would later search for the ship's debris in Texas and eventually run the *Columbia* Preservation Office. *(Photo courtesy Mike Ciannilli)*

An observer in Firing Room 4 of the Launch Control Center watches *Columbia* roll out to the launch pad on December 9, 2002. *(NASA photo)*

The STS-107 crew poses with Robert Hanley (kneeling) at the completion of the terminal countdown demonstration test, December 2002. Crew (left to right): Mike Anderson, Rick Husband, Laurel Clark, Willie McCool, Ilan Ramon, Kalpana Chawla, Dave Brown.
*(Photo courtesy Robert Hanley)*

The STS-107 crew leaves the Operatio
and Checkout Building en route to the
launch pad, January 16, 2003.
*(Scott Andrews/NASA photo)*

"Kid pic" drawn by the children of *Columbia*'s crew in the Launch Control Center in the hours prior to liftoff. *(Jonathan Ward photo)*

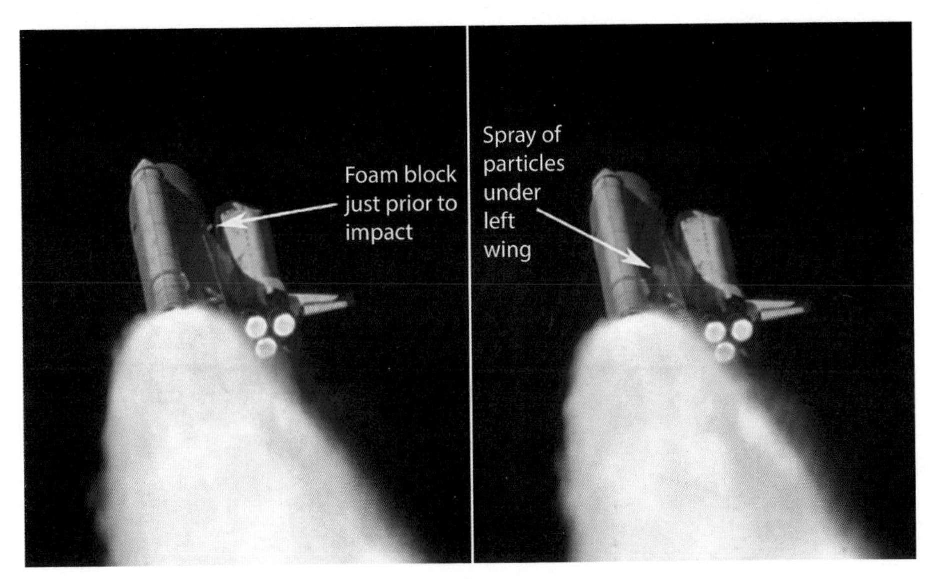


Ascent film frames of *Columbia* just before and after the impact with the chunk of insulating foam that fell from its external tank. *(NASA photo)*

The space shuttle's main design weakness was that the orbiter was susceptible to impacts from debris falling off the external tank or the solid rocket boosters. With STS-107, a piece of foam fell off the external tank's bipod ramp and struck the orbiter's left wing during ascent to orbit. *(NASA photo)*

The status board at Mission Control at 9:15 a.m. EST, when *Columbia* should have been lining up for landing at Kennedy. The highlighted numbers are "stale data"—information not updated since last contact with the ship at 9:00 a.m. EST. Flight controllers were unaware that *Columbia* had already disintegrated. *(NASA photo)*

The Etoile Volunteer Fire Department investigates a piece of debris along State Route 103 the morning of the accident. *(Jan Amen photo)*

Volunteers traverse the bayous of East Texas looking for remains of *Columbia*'s crew. *(Tom Iraci/US Forest Service)*

The truck bays of Hemphill's Volunteer Fire Department became the area's Incident Command Post. *(Jan Amen photo)*

A cake decorated with the names of *Columbia*'s crew, donated by an anonymous Sabine County volunteer. *(Jan Amen photo)*

Troopers from the Texas Department of Public Safety grab an early dinner at the Hemphill VFW. *(Jan Amen photo)*

The Ladies Auxiliary meeting room at the Hemphill VFW was pressed into service to store some of the food and goods donated by the local community. *(Jan Amen photo)*

The recently completed memorial star in the center of Hemphill became a place for local residents to leave tokens of sympathy for the crew and NASA after the accident. *(Jan Amen photo)*

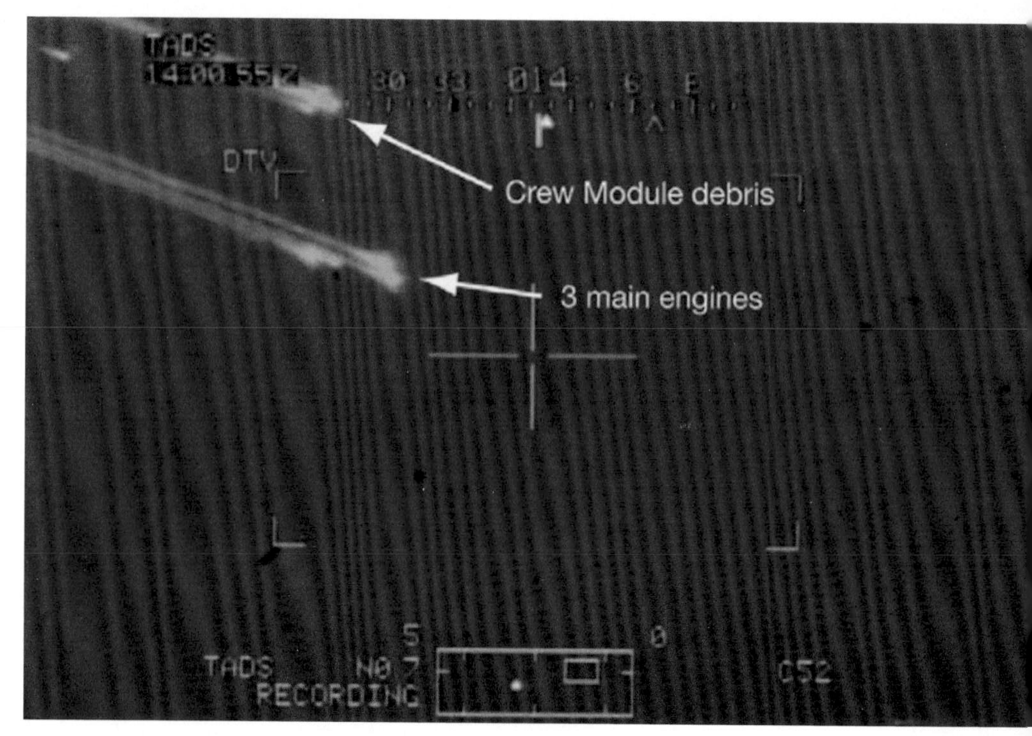


Annotated frame from a gunsight camera video of *Columbia*'s breakup, taken by the pilot of an Apache helicopter out of Fort Hood, Texas. The detailed information recorded by this camera system was vital to the investigation of the accident. *(NASA photo)*

Debris at Barksdale Air Force Base awaits transport to Kennedy Space Center. Several of *Columbia*'s propellant tanks are at right rear. Pieces of the airlock and Spacehab tunnel structure are at lower right. At center is part of the right landing gear door. This photo was taken on February 8, 2003—one week after the accident. *(NASA photo)*

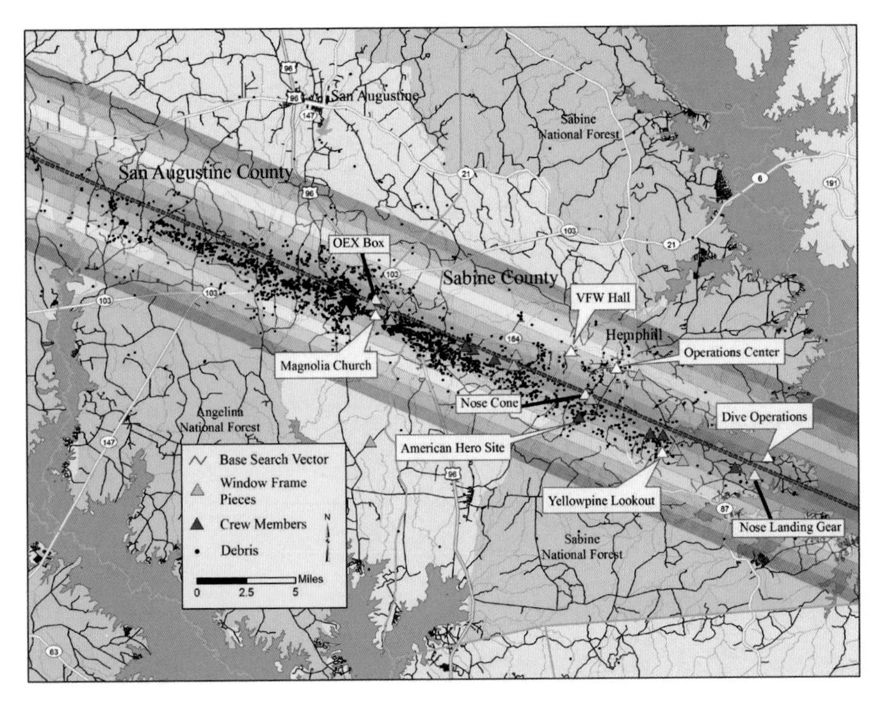


Map of some of the key recoveries in San Augustine and Sabine Counties. The black dots indicate debris found and tagged within the first two weeks following the accident. *(Jeff Williams/Stephen F. Austin State University)*

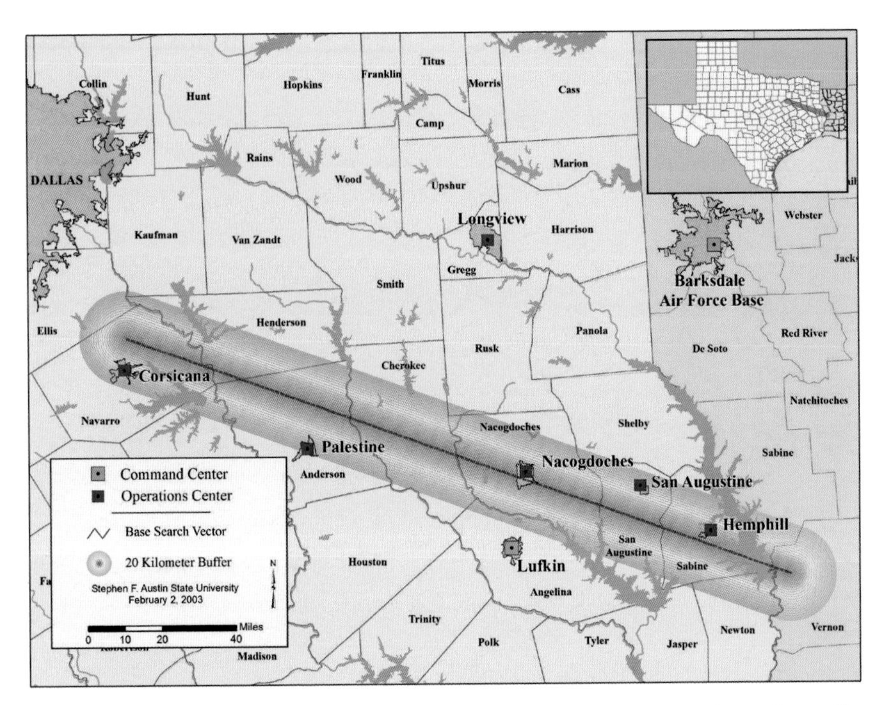


About 95 percent of *Columbia*'s debris came down within an area about five miles on either side of the base search vector—the path followed by the main engines after the vehicle broke up. *(Jeff Williams/Stephen F. Austin State University)*

One of the "tent cities" for fire crews, this one in a warehouse that served as the Incident Command Center in Palestine, Texas. *(Jan Amen photo)*

A fire crew grid searches an open field on a sunny day in East Texas. Between mid-February and late April, hundreds of teams like this one were deployed every day. *(Tom Iraci/US Forest Service)*

The boot-drying tent at the Nacogdoches command post. FEMA provided a second pair of boots to every searcher at no cost, enabling fire crew members to wear dry footgear every day. *(Tom Iraci/US Forest Service)*

This cartoon, drawn by a Navajo firefighter from Arizona, depicts the hazards and rewards of searching for *Columbia* debris.

*(Courtesy Patricia Huffman Smith "Remembering* Columbia*" Museum)*

A searcher ducks under a barbed wire fence near Hemphill. Searchers had to go over or under obstacles—not around them—to ensure nothing was missed. *(Jan Amen photo)*

The Navy's search efforts on Toledo Bend Reservoir proved frustrating, as no significant debris was recovered despite more than three thousand dives. *(Jan Amen photo)*

Pat Adkins (standing in truck bed) hoses mud off of *Columbia*'s nose landing gear, recovered on February 18, 2003. *(NASA photo)*

A muddy crater where one of *Columbia*'s turbopumps slammed into the ground at Fort Polk, Louisiana. *(NASA photo)*

One of *Columbia*'s main engine powerheads, pulled out from under fourteen feet of Louisiana mud. *(NASA photo)*

Workers inspect *Columbia*'s #2 main engine before the STS-107 mission. The powerhead is at left in this NASA photo.

A searcher in one of the National Guard units suffered an injury. The unit became lost when their GPS device failed, and they called for assistance. Terry Lane and Cohrs located them from their previous GPS report. Cohrs learned that the troops had not been issued compasses. After a few calls from Cohrs and the DPS, the guardsmen had compasses the next morning as backup to their GPS devices.

After a week in the field, team leader Jamie Sowell was tired and grumpy. Three elderly volunteers were among his new team members this Saturday morning. Sowell was frustrated to have to continually rehash the basics of search protocol. Time was too valuable to spend half a day bringing new people up to speed. He issued terse instructions on how to conduct the search and then spread his team out in a line through the woods.

Team leaders probably walked three times the distance of the other searchers every day. They constantly walked back and forth behind the lines, issuing instructions to keep searchers evenly spaced and in a straight line so the area could be completely searched.

In "cold" areas, there were times when debris sightings were few and far between. Searchers wondered if they were overlooking things if too much time passed between sightings. When someone found a piece of debris, curiosity naturally got the better of many of the volunteers, and they wanted to see for themselves what the debris looked like. With fifty people spaced ten feet apart from one another through the dense undergrowth, chaos quickly ensued when people gathered to examine a newly found object. It could take an hour to get everyone back into position and resume the search.

After Sowell's crew found its first object of the day, his three new elderly volunteers broke from their positions to examine it. They argued over whether the object was a piece of the shuttle, because it appeared to have rust on it. Sowell began to lose his temper. "Look, guys, we're not a bunch of rocket scientists. You need to get back into line!" To his surprise, one of the men told him that, in fact, they *were* rocket scientists—retired Apollo-era NASA employees. Sowell was briefly speechless. Regaining his composure, he said, "That's nice. But get back in line!"

After a week on-site in Hemphill, NASA's workers from Kennedy continued to be overwhelmed by the local citizens' acts of charity and kindness. NASA workers found that their money was no good in town—the grocery store refused to let them pay. In return, the NASA personnel freely offered mission pins, patches, and crew photos, which the townspeople eagerly received.

Wet conditions in the field soaked the searchers' socks and boots, rendering the cold intolerable. There was so little time between searches that hanging socks up to dry was not enough. Hearing that, one of the senior loan officers from Shelby Savings Bank purchased for the searchers all the socks available in Hemphill's stores, and also bought all that were available in the Walmart at Jasper.

Pat Oden emailed a friend: "We have more National Guard that arrived Thursday, too. They are very young, and when you meet them in town, at the grocery store, etc., they come up to you and thank you for taking such good care of them. They are being housed mainly at the school gym, and have been taken more blankets, pillows, etc. We are now washing the searchers' clothes, as the going has been tough—wet and cold . . . All, to a man, thank us profusely for our help and concern for their well being during this emotional task. And it is very emotional . . . my neighbors have been searching also, and two of them and their team found human remains day before yesterday. The man staying with me found remains yesterday. Most of the remains are being found in and near Hemphill, most in our dense national forests, briar patches up to six feet tall and they must go through these, not around them. Very hard on these men, but they are very dedicated and rising to the occasion."[32]

The tight bond of the small town to the recovery of *Columbia* and her crew led to a saying that quickly spread throughout the community: "Their mission became our mission."[33]

—

The following day, Sunday, February 9, 1,350 searchers walked in the areas of Sabine County between US 96 and Toledo Bend Reservoir, trying to complete the coverage of a mile-wide path along the new search centerline. The deployment went quicker than the previous day, but there were still problems coordinating so many people. Searchers worried that the cold rain would cause them to be pulled from the field again.

Overnight, the Lufkin command staff decided to pull one of the National Guard teams from Hemphill to search in an area of interest in San Augustine. However, they forgot to inform the Hemphill command post that the unit was being redeployed, which created accountability concerns.

Greg Cohrs felt obligated to accept all the volunteer help being offered, but the huge influx of searchers was overwhelming the leadership span of control and the incident support system. As he feared, the inefficiency of working with so many untrained people in large groups meant that not much debris was located over the weekend. More troublesome, the two remaining crew members were still missing.

After a period of high winds damaged branches throughout the forest, helicopter searches for broken branches in the treetops were no longer producing results. Instead, Don Eddings, flying in a US Forest Service–contracted helicopter, was now on the lookout for buzzards. Every mid-morning, the birds would leave their rooftop perches and begin circling. The search team's helicopter would fly in to scatter the buzzards and then notice where they started circling again. Eddings would call in the location so ground teams could search the area.

Airborne assets from NASA and the Department of Defense also aided in the search. Although specific details on the technologies involved were not divulged, the incident command team could get detailed information on the location, movement, and number of coyotes and buzzards in the forest.

—

After so many days of frustrating searches without results, Greg Cohrs and Terry Lane each prayed on the morning of February 10 that the search for the crew would be fruitful that day.[34] During Monday's morning briefing at the VFW hall, Brother Fred Raney led a prayer for good weather. He closed with, "Lord, let us find the remaining astronauts."

Help from the Department of Interior arrived. Cohrs was transitioned to another incident command team, led by Mark Ruggiero of the US Fish and Wildlife Service. Relieved from having to attend so many meetings, Cohrs could now better focus on planning and directing the search operations.

About 850 searchers were in Sabine County that day, down from the weekend's mob. To everyone's surprise and frustration, the contingent from the Department of Public Safety—about three hundred officers—had been rotated out and replaced by an entirely new group, who had to be brought up to speed. School was back in session, which complicated getting the searchers to and from the field, since they had been using school buses over the weekend.

Feedback from the search team leaders led Cohrs to believe that the area between Highway 87 and the Yellowpine Lookout tower had not been covered effectively. He sent some of the searchers back to cover the area again. He had a strong hunch that they might find at least one crew member there. They found nothing of significance.

One of the National Guard units, however, located one of the missing astronauts in mid-morning near Toledo Bend Reservoir and Farm Road 2928. Everyone was elated that another crew member had been found after nine days of such intense effort. However, since this sighting was so close to the reservoir, Cohrs was worried that the last crew member might be in the lake.

Cohrs directed Don Eddings to search the Six Mile and Big Sandy Coves areas of Toledo Bend. Eddings saw nothing from his helicopter.

Rumors spread among the search teams that the last of the crew members had finally been recovered. There was an understandable and unfortunate letdown when information came that evening that the seventh crew member had yet to be located.[35]

During his search team's lunch break, local resident Mike Alexander struck up a conversation with Dan Sauerwein, a volunteer searcher who was new to the team that day. Sauerwein told Alexander that he worked in the Neutral Buoyancy Lab—the enormous pool used for space walk training in Houston. Sauerwein took a week's leave from his job and drove to Hemphill to help search for the astronauts. He brought sturdy clothing but had neglected to make arrangements for lodging, so he had slept the previous night in his car. Alexander immediately offered him a room at his house, joining two other searchers that Alexander was already hosting. Sauerwein later said that he had never been treated so well. As happened countless other times throughout the recovery period, this act of kindness between unlikely strangers cemented a long-lasting friendship.

Greg Cohrs's day did not end on a happy note, despite the recovery of the sixth member of *Columbia*'s crew. The command trailer in which he was working was beset by an infestation of wasps.[36]

—

On Tuesday, February 11, Mike Alexander and Dan Sauerwein's search team emerged from the woods of the Sabine National Forest in sight of the Toledo Bend Reservoir. Someone looked behind them and called for the line to stop. Turning around, Sauerwein was amazed to see shredded canvas high in the treetops—material from one of *Columbia*'s experiment packages. He thought that if so much of that material was in the treetops near the reservoir, there must be even more in the lake.

In the mid-morning, one of the Forest Service teams called to report they had located the last crew member, near Housen Bayou between Route 87 and the Yellowpine Lookout. This was in the vicinity of where the FBI and Greg Cohrs had suspected something significant would be found.

Cohrs's feelings of joy and relief overwhelmed him. Successfully recovering the crew had seemed so improbable ten days ago, considering that *Columbia* had disintegrated two hundred thousand feet high and traveling in excess of 12,000 miles per hour. His careful planning, his insistence on maintaining a disciplined and methodical search, his determination, and his faith had all paid off. All of the crew members' remains were recovered along a fourteen-mile-long path within the one-mile-wide search area.

The crew members would all be going home.

Word about the final recovery came by radio to the other search teams in the field. When he heard the announcement, Mike Alexander broke into tears. "I just started crying out there in the woods. I couldn't help it. I thought, *We got closure now*."

Don Eddings returned to the command post from his air search that day to find people shaking hands and congratulating one another. He asked the ranger what was going on. "Didn't they tell you? We found the last one today. It was one of the locations you called in."

In the afternoon, Cohrs suggested to Brent Jett that Hemphill hold a memorial service for the crew. All of the Sabine County search teams had been in the field when the televised memorial services were held at other locations. The community wanted to pay its respects to *Columbia*'s crew after working so hard to bring them home.

That evening, Jett asked Cohrs to walk with him to his car. The streets were already empty, the media having moved on several days ago to cover other stories. Jett told Cohrs how deeply he and NASA appreciated the work of Brother Fred Raney, Terry Lane, and Cohrs in helping to search for and recover his friends and members of the NASA family.

—

NASA's Dave King knew his leadership team needed a break after twelve consecutive twenty-hour workdays, so on Wednesday he took them to dinner at an Italian restaurant near the Lufkin command center.

As they ate, King noticed a young girl looking at them from an adjacent table. She came over a few minutes later and asked, "Are you

NASA people, and are you trying to recover the space shuttle? I just want to tell you how much I appreciate what you people have done for the space program and our country."

He found it particularly poignant that this young girl had come over to them and said the words they most needed to hear at precisely the right moment. "That's exactly what encourages you—the human piece of this thing," he later said. "To have this little girl, who you look to as the future of this country, to come up and say that . . . it was unbelievably meaningful to us in that moment. We had all worked very hard and were all very tired, but something like that gives you a new shot of energy to go do what you need to do."

—

Thursday, February 13, was the final day of the Sabine County incident management team's formal leadership of the search efforts in the county. Responsibility would transition to a new leadership structure on Sunday, February 16.

In the meantime, the Hemphill camp worked a short day on Thursday, then took a well-deserved break to reflect on the events of the past two weeks. Search operations paused on Friday to memorialize the *Columbia* crew.

Mary Beth Gray of Hemphill's flower shop prepared red, white, and blue wreaths for *Columbia*'s crew members. Friday morning's memorial service began at eight o'clock in the VFW hall, with a procession of people carrying the seven wreaths to the front of the room. Members of the incident command center spoke briefly. Brent Jett delivered the eulogy for his colleagues. He thanked the community for their efforts in bringing *Columbia*'s crew out of the woods and back to their loved ones.

Minimal response to calls resumed on Saturday. The teams had not yet completed searching every square foot of the twenty-five-square-mile target area. Partial remains, somehow missed in one of the earlier searches, turned up over the weekend. But except for one small find on March 7—outside of the search area—those were the last remains of *Columbia*'s crew to be located.

—

Sabine County's citizens paused to reflect on the profound emotional and logistical challenges of the past two weeks.

It was impossible to determine precisely how much food had been donated or how many meals had been served at Hemphill's VFW hall. Best estimates were that this relatively small community and a handful of volunteers had prepared and served thirty thousand to fifty thousand meals in two weeks. The community had donated more than $620,000 in services to the recovery—at no cost to the federal or state government.

No doubt Sabine County's sense of personal ownership and connection to the search was in large part due to the crew's coming to rest in the county. Had this been simply a hardware salvage operation, the local citizens might have been interested—but perhaps not so compelled to go to the lengths they did to help. But the presence of the crew and their worried and grieving friends made this a tragedy that touched every compassionate fiber of their being.

Many people in this deeply religious area saw it as divine providence that *Columbia*'s crew came to Earth in their community. Had the accident occurred anywhere else, the outcome might have been very different. Here, the people kept what they saw and experienced out of the press. They enfolded the NASA family in a respectful, loving, and healing embrace, and they rose to address a national tragedy in a manner that is difficult for outsiders to fully comprehend.

"Texans are really just good people and will give you the shirt off their back," said Jan Amen. "We might get a bad rap with this redneck thing, but we're just down home good people."

Every NASA employee and every person who participated in the recovery spoke of the miraculous, loving deeds performed by Sabine County's citizens. Even the neighboring communities were astounded by the way the town rose to the challenge. Perhaps it was just the nature of small-town existence. Many residents played multiple roles as part of their everyday life, such as Brother Fred Raney being both the Baptist minister and captain of the volunteer fire department. In larger towns like Nacogdoches, the culture was such that it seemed more appropriate to have the formal organizations and their professional resources deal with the incident.

"It's just people helping people—that's what this small town is about," said Roger Gay. "Everybody likes to help everybody else, and they don't expect anything from it. It was an occurrence that happened, and we dealt with it the best way we knew how."

NASA's Dave King summed up, "The people of East Texas make you proud to be an American, because they sacrificed and gave everything they had to try to help us. It was unbelievable what they did for us."

# PART III

# PICKING UP THE PIECES

These really rough, hard-core, no-nonsense, work-hard people on the Native American fire crews would treat every piece they found with such reverence. It wasn't an inanimate object; it was a very animate, very personal thing. They understand that everything around us is a living, breathing being that we cooperate with. It made me appreciate my heritage, what these people sacrifice, and how special this experience was to them.

*—John Herrington, the first Native American astronaut*

## Chapter 8

# *COLUMBIA* IS GOING HOME IN A COFFIN

By the third day into the search period, while the crew recovery teams were desperately looking for *Columbia*'s astronauts, we on the Mishap Investigation Team in Barksdale and Lufkin were still trying to get a handle on the size of the debris situation. I was acting as Dave Whittle's right-hand man at this point, doing whatever I could to support him and manage the interfaces within NASA and with the other agencies. An incredible amount of activity was underway.

At the nine o'clock morning meeting on Monday, February 3, at Barksdale, Paul Monafo from Marshall Space Flight Center reported he was developing a fault tree, which was an analysis of how different basic events or failures might have combined to cause the destruction of *Columbia*. The NTSB urged people to keep an open mind and not discount any possibilities, no matter how improbable. Monafo, like many NASA managers, nonetheless believed the accident could not have been caused just by the foam impact during ascent. Many NASA managers held onto this opinion for days—even weeks—after the accident. It seemed impossible that a piece of lightweight foam by itself could have damaged the shuttle. *There must have been something else involved.*

We also began discussing how we might reconstruct *Columbia*'s debris to determine the cause of the accident. Representatives of both Johnson Space Center and Kennedy Space Center made the case for why their center should be the location for this important investigation. Ron Dittemore and representatives from Johnson Space Center lobbied to bring the debris to Houston for analysis. Johnson had the advantage of being close to the debris field and was home to the engineers with design knowledge of *Columbia.* On the other hand, Kennedy had an available hangar at the Shuttle Landing Facility that would be ideal for laying out *Columbia*'s wreckage in a reconstruction process similar to that used by the NTSB for investigating aircraft accidents. More important, we said, our technicians were more intimately familiar with the flight hardware than were JSC's engineers.

Fred Gregory, NASA deputy administrator and former shuttle commander, was in Barksdale for several days. I discussed with him the pros and cons of the two options. Not knowing at the time that I would later lead the reconstruction effort, I told Fred that I firmly believed that KSC was the better choice for the effort and asked him to support my suggestion.

I asked KSC's safety manager Gerry Schumann to go to Lufkin to support Dave King. King wanted him to coordinate safety briefings for the press and also to ensure that people working in the field understood risks and how to avoid exposure to hazards. Everyone participating in the searches had to be briefed on the hazardous materials on the shuttle, as well as the other risks they might encounter in the field. The news would get out through the joint information center established at Lufkin—a clearinghouse for all the federal agencies on site to disseminate information related to the disaster.[1]

We had quickly determined it was better to have Jerry Ross and Ed Mango run the debris collection process locally in Lufkin—close to the debris field—rather than remotely from Barksdale. Ross and Mango continued to set up our debris collection protocol. They ensured the local collection centers were on the lookout for items of particular importance. We hoped that at least one of *Columbia*'s five onboard flight computers might have made it to the ground. The computers had limited battery power to retain data in memory, and we wanted to recover any data that might still be in the computers before the battery backup power failed. Hazardous items also needed to be removed from the field immediately.

Limited resources in the first several days of the operation meant that most debris items could only be marked, their GPS coordinates recorded, and then left in the field for collection later. If an EPA representative was on a search team, that person could examine an item in the field for hazardous substances and approve it to be picked up if it was not contaminated. We planned to deploy twenty-four NASA teams to the field to assist with debris collection.

Ross directed that any of the crew's personal equipment items located during the searches be sent directly to him for examination. Over the course of the coming weeks, he saw kneeboards, checklists, and pieces of the flight suits from *Columbia*'s crew.

At the collection and storage sites, there were not enough NASA resources to perform triage on the material coming in from the field. I called Denny Gagen at KSC and asked that he send more people to Texas as quickly as possible. In the meantime, collection centers had instructions to check each item for hazards, enter its brief description and GPS coordinates in the database, and then ship it on to Lufkin. From there, it would go by truck to Barksdale. Betty Muldowney—a quality control manager from KSC who came out with us on the first flight—was organizing the process for receiving debris at Barksdale, logging it, and transferring it to wherever it would be studied.

—

At 12:30, Administrator O'Keefe and Fred Gregory declared that the hangar at the southeast end of Kennedy's Shuttle Landing Facility would be the location for examining *Columbia*'s debris. Recovered parts of the shuttle were to be shipped there from Barksdale as soon as the hangar was ready.

We heard reports from California of three possible pieces of shuttle debris. In addition, someone reported a debris field near Phoenix.

If these sightings checked out, we would dispatch recovery teams to both locations.

Our calls for more resources were yielding results, and the magnitude of the nation's response to NASA's need was incredible.[2] An additional sixty people would be deployed from Kennedy by Thursday to help with debris recovery and processing. Ralph Roe said that he would be sending people from Houston to Barksdale to help identify "interesting" material. Later in the afternoon, Houston also committed to sending three teams to Lufkin to join the debris collection and identification effort. A twenty-five-member disaster mortuary team was headed to Lufkin to assist in working with remains of *Columbia*'s crew.[3] The FBI was deploying Evidence Response Team members to ensure that debris collection practices were consistent across search teams and collection sites. The Texas National Guard announced that they would deploy 477 troops to help with debris collection beginning the next day. The coast guard sent additional personnel from its Atlantic, Pacific, and Gulf Strike Teams to Lufkin. FEMA was establishing a third disaster field office at the Fort Worth Naval Air Station. And the EPA was analyzing water samples in Louisiana to check for contamination by shuttle debris.

Astronaut Jim Reilly visited Toledo Bend Reservoir with representatives from EPA, FEMA, and the coast guard. They searched part of the shoreline, near where fishermen had reported seeing objects impact the water, but they found nothing.[4] Jasper sheriff Billy Rowles was managing the local water search effort. Boats and dive teams from the local law enforcement and the FBI operated out of the Fin and Feather Resort attempting to locate debris. Reilly reported that the water search effort needed to be expanded.

Meanwhile, other welcome resources appeared on the scene. The Blue Bell Company provided ice cream coolers at all of the command centers and field offices and kept them stocked, at no cost to the recovery operation. Community Coffee provided coffee machines at every one of the collection centers, also at no cost. The Salvation Army also provided meals for all of the personnel in the Lufkin command center.[5]

—

Nathan Ener and Tim "Peewee" Mitchell, residents of Bayou Bend Road west of Hemphill, spotted broken treetops in the forest canopy near Ener's house.[6] Mitchell saw what he thought to be a small garbage dump in the forest floor below. They examined the site more closely. Lying in a shallow crater on the forest floor below was *Columbia*'s nose cap and some of its supporting structure (which some local citizens referred to as the "nose cone"). The dark gray reinforced carbon-carbon dome was cracked from colliding with the trees at high speed. Ener called in his find, but his description was vague, and the people taking the call did not immediately understand its importance.[7]

It is understandable why the call might have been overlooked. By the end of the day on Monday—only two and one half days after the accident—there were tens of thousands of reports of debris on the ground in thirty-three counties in Texas. About twelve thousand pieces of debris had already been collected.

The lack of communications infrastructure was one of the most frustrating bottlenecks early in the accident investigation, especially when trying to talk to people in the field. Cell phone service was practically nonexistent. Calls to Hemphill had to be routed through the fire station's front desk, which was overwhelmed with the calls coming in from searchers in the field and from local residents who were still finding material on their property. Calling back from the field to Lufkin was also challenging. Gerry Schumann resorted to making his calls on the pay phone at a Hemphill grocery store.

Although NASA and FEMA had provided satellite phones to some staff, they were difficult to use. Callers needed to be on high, unobstructed ground to get a clear shot at the satellite—impossible in Sabine County's dense forests. Even in more wide-open places like Nacogdoches or Palestine, a satellite phone user had to stand out in the middle of a field or on top of a hill to make a call. The phones' batteries ran down quickly.

Our temporary work-around was a network of runners. Each morning's status meeting in Lufkin concluded with written instructions for each of the recovery sites. Then runners would drive the written messages to the leaders at each of the debris collection centers. If there was any news to go back to Lufkin, the runners could bring that information back by car.

Our teams used the runner system for the rest of the first week of the debris recovery operation. The process seems laughably quaint or inefficient now, especially for a high-tech organization such as NASA. However, it was the only solution at the time for exchanging detailed information in remote, rural locations where landlines, cell phones, satellite phones, and high-speed Internet were not widely available.

Meanwhile, cleanup of debris around schools, hospitals, and other public places continued. FEMA worked with the county judges in each of the affected counties to determine priorities for clearing debris. The county judges and FEMA separately reported back to Governor Perry about the status of the process. Good progress was being made by the end of the day, with about half of the schools reported free from *Columbia* debris.

—

KSC quality control inspector Pat Adkins drove from Shreveport to Lufkin to get his assignment for the debris recovery operations. Because of his familiarity with *Columbia*'s crew module, he deployed to the San Augustine and Hemphill areas, where items from that section of the shuttle were being found. On the drive through the countryside toward Hemphill, Adkins noticed places where the road and shoulders were dusted with something that looked like fine snow. It turned out to be a powder composed of small chunks of tile and silica from the shuttle's heatshield.[8]

Adkins arrived in the early evening at the San Augustine command center, which was essentially an old house that had been converted to a community meeting hall. He recognized astronaut Chris Ferguson, who was sitting at a table and cataloging some recovered items into a spreadsheet on his laptop. Ferguson pointed to the closet and told Adkins that it contained plastic bags filled with items that had not been examined yet. Adkins looked inside and immediately smelled oxidizer from *Columbia*'s propulsion system. He propped open the door and a window to let the room air out.

Adkins donned latex gloves and began going through the material, separating out the items that might have been contaminated with toxic propellants. Everything was wet and muddy. In the bottom of the first bag, he saw a wristwatch with a fogged-over blue face. A few minutes later, he found a music CD with Hebrew writing on it. These startling objects were a grim reminder of what had occurred—a wake-up call for the emotional challenges he would be facing in the coming days and weeks. He took the crew's personal items out of the closet, washed them off, bagged them, and gave them to Ferguson.[9]

—

Thanks to an all-out effort by the EPA, Texas Commission on Environmental Quality, local resources, and NASA, all of the school grounds along the debris path in Texas were reported clear of shuttle debris by late evening on February 3. Schools opened and students returned to their classes Tuesday morning.

Meanwhile, more resources were showing up to recover and process shuttle debris. By February 4, the EPA had thirty-eight debris recovery teams on the ground in Texas, with an additional eighty teams expected by the end of the week. One hundred members of the FBI's Evidence Response Teams were on the ground.[10] The C-141 cargo plane scheduled to bring another wave of our NASA technicians to Barksdale from Kennedy was unable to make the flight Monday night because of mechanical problems, but a replacement aircraft was due Tuesday morning.

Although we were making some headway, the scope of the debris problem continued to grow. Half of Louisiana's sites had been cleared of debris,[11] but the overall land area that needed to be searched thoroughly continued to expand. Thirty-eight Texas counties were now reporting debris on the ground, with sightings of possible debris in California, Arizona, and Nevada. NASA and the EPA agreed to investigate those sightings.[12] FEMA itself had no authority under the disaster declaration to enter those states unless shuttle debris was actually confirmed there.[13]

A US Coast Guard strike team went out onto Toledo Bend Reservoir on this cold, rainy morning. The twenty-four-foot boat with side-scan sonar and divers would supplement the dive teams already on-site searching for debris.[14]

—

Nathan Ener[15] once again phoned the Hemphill command center to remind them about finding *Columbia*'s nose cap the previous day. This time, he described the object in enough detail to attract immediate attention from the people taking the call. A NASA team came out to the site and confirmed that this was clearly a significant piece of the shuttle.[16]

NASA permitted the media to cover the process of extracting the nose cap from the woods—one of the largest operations related to a single piece of debris in Sabine County. The state police blocked off Bayou Bend Road, requiring the media to park along Route 83. Marsha Cooper from the US Forest Service was assigned to take the media to the site. She deliberately led them on a circuitous route to confuse them and prevent them from easily finding their way back unescorted.

The original extraction plan called for lifting out the bulky piece by helicopter. Felix Holmes of the US Forest Service cut down some of the large trees surrounding the wreckage to give a helicopter a clear shot at lifting the nose cap and its supporting pallet out through the forest canopy. The weather would not cooperate, though, and a strong cold front and rain blew in shortly before the scheduled operation.

The flight was canceled. It was too dangerous to risk the helicopter and the priceless piece of the shuttle by attempting to hoist the heavy object up through the trees in a high wind. Holmes cleared a path so that a four-wheeler towing a trailer could haul the nose cap out to a waiting tractor trailer for transport to Barksdale.

Meanwhile, Pat Adkins investigated a call reporting a spherical tank resting against a fence line. He realized immediately that it was still "hot"—its lethal contents of nitrogen tetroxide gradually seeping out and fuming upon interaction with the rain. The tank was too hazardous to approach, and the ground around it was contaminated. He placed a "crime scene" tape at a safe distance around the tank. He asked a Texas DPS trooper to keep watch over it—and stay upwind of the tank—until the EPA could decontaminate it and collect it.

NTSB investigator Clint Crookshanks was with a search team that found about one hundred pages of a manual in what had once been a three-ring binder. The cover of the binder was missing, but the pages appeared almost untouched and unburned.[17] Other finds were not so well preserved. Once-pristine metallic components were now heavily oxidized, twisted, and scorched. Some of the debris resembled car parts that had been rusting in a junkyard for fifty years.

Many of the hundreds of pyrotechnic and pressure devices that had been aboard *Columbia* turned up along the debris path. Some were large, while others were as small as BBs—like the initiators for the inflation cartridges in the crew's life vests. Everything had to be treated as hazardous, because they may not have been expended and might be damaged, which could cause them to go off without warning.

—

The various collection sites along the debris corridor checked in at six o'clock that evening. Ed Mango, Dom Gorie, and Jerry Ross reported from Lufkin that "a lot" of material was being collected. Teams had identified helium and gaseous nitrogen tanks in the field but had not picked them up yet. Some 70mm film had also been recovered. As teams filled their vehicles with debris, they drove the material to the nearest collection center for processing. Once a collection center received enough material to fill a semitrailer, the packaged items would be sent to us at Barksdale.

At the end of February 3, we had asked the debris collection centers to note how many items had been recovered that day, and, if possible, to describe what they might be. The next day, Lufkin requested specific GPS locations for the items collected. Several times each day over the next several days, the debris team in Lufkin relayed the information by phone to us at Barksdale, with verbal descriptions of significant items that had been recovered.

The problem was that each collection center—and in many cases each agency—was recording debris sightings and recovery in their own way. In the opening days of the recovery operation, information was being handwritten on forms and faxed to Lufkin. Sites with computers logged the debris with software ranging from makeshift Microsoft Excel spreadsheets to more sophisticated databases. The GPS locations of debris sightings—and even of crew remains—were being recorded inconsistently in the field, with some teams recording decimal degrees and others using degrees, minutes, and seconds, resulting in misunderstandings on the ground. To add to the confusion, the same piece of debris sometimes appeared on multiple databases identified with different key numbers.[18]

It was impossible for us to get consistent answers to basic questions, such as which items had been recovered and where, where they were being stored, whether the debris had been identified, and when it would be shipped to the processing site at Barksdale. It was particularly important for NASA to get firm answers on where items from the crew module were being recovered, so that teams searching for the crew remains could be more precisely deployed.

Database management and reconciliation was quickly becoming a nightmare. With tens of thousands of pieces of debris being found, we could soon lose control of the situation.

In response, FEMA created and fielded a new Shuttle Interagency Debris Database using the Geographical Information System, which could help generate search maps. The collection centers initially emailed back to Lufkin a spreadsheet at the end of each day. Later, a NASA team from Houston created a web-based method to simplify data input and to help document and validate the data. But while this effort was underway, the EPA continued to collect debris in the field and enter the information into its own database.[19]

This challenge was resolved by working out how to pass information back and forth between FEMA's and the EPA's databases. An interagency team spent weeks working through the tens of thousands of records already entered to identify duplicate records and enter all the information into the new format. This was critical for targeting searches later in the recovery operation.

One of the most important lessons learned from the debris recovery operation was the need to have an agreed format for databases to use in all-incident emergencies.[20]

—

At our request, the Texas Department of Public Safety issued guidelines on February 5 for how first responders should deal with shuttle debris. Private citizens were told not to pick up any debris, even though much of it was likely not hazardous. First responders needed to be aware of dangers that included: stored energy (high-pressure tanks and cylinders); monomethyl hydrazine, nitrogen tetroxide, and ammonia; pyrotechnic devices (anything marked yellow/black near window frames, landing gear, crew seats, hatches, and antennae); and biological material.[21]

Responders were ordered to stay with anything marked SECRET, CONFIDENTIAL, or SSOR (space-to-space orbiter radio) until someone from NASA arrived on the scene to collect it personally.[22]

NASA requested that FEMA alert the residents of seven rural counties in Texas west of Fort Worth to be on the lookout for possible scattered shuttle material. Data analysis suggested that shuttle debris might have fallen farther west than we previously thought.[23] We continued to refine the expected debris path using NASA's ground track for the shuttle above eighty thousand feet and Department of Defense radar data below eighty thousand feet.

—

The FBI said they were investigating approximately twenty reported thefts of shuttle wreckage. They were also looking into seventeen Internet auctions of what people claimed to be pieces of *Columbia*. Officials suspected that souvenir hunters illegally collected over one hundred pieces of the shuttle.[24]

The US Attorneys Office announced a limited prosecution moratorium—until Friday, February 7, at five o'clock—for people who voluntarily turned in shuttle debris. Other callers returned items that they had picked up but did not initially turn over for fear of being accused of tampering with evidence.

Interestingly, after the moratorium was announced for *Columbia* debris, NASA received a few calls asking if the moratorium also applied to material from the *Challenger* accident seventeen years earlier. NASA said yes—and several pieces of *Challenger*'s wreckage were then turned in.[25]

—

By February 5, Humanities Undergraduate Environmental Sciences (HUES) and Forest Resources Institute labs at Stephen F. Austin State University in Nacogdoches were printing more than one thousand search maps every day. The university also fielded nearly seventy search teams, with almost two hundred people working to locate shuttle debris. Their state-of-the-art GPS equipment and processing software allowed debris locations to be tagged within an accuracy of three feet. Recovery leaders were so impressed with the capability that they installed a high-speed data line between the university and the command center so that the maps could be printed in Lufkin.[26]

After every evening's briefing for the Lufkin command staff, the Forest Resources Institute printed out geodetic maps for the next day's search area. The maps showed the search objectives printed on aerial and satellite photographs of the area two miles on either side of the centerline computed by Jerry Ross and John Grunsfeld. Then the ground bosses delivered the maps to each of the forward search centers.

Translating the maps into action entailed briefing the county judges about the designated search locations, planning the logistics for deploying the searchers to the field, and notifying landowners in the search path. Search team leaders needed to know if there were territorial bulls in pastures that they would be crossing.

The day's other good news came in the form of a plane from Kennedy that arrived at Barksdale at eleven o'clock. It brought us another sixty staff to assist in the debris recovery and processing effort.

—

On Thursday, February 6, at our morning meeting in Barksdale, we first heard about the "Day 2 object" (described in chapter 3). It was intriguing—and disturbing—to learn something had been in orbit with *Columbia* early in its mission.

With the NTSB's guidance, NASA's internal accident investigation was taking two independent tracks. The data team in Houston concentrated on the telemetry and other data from the shuttle to determine the failure sequence. Our debris team would examine the physical evidence of how the shuttle likely came apart. Each team would develop their conclusions independently. If the two sets of conclusions matched, then NASA could be certain that it understood the cause of the accident.

We discussed the authority of the *Columbia* Accident Investigation Board and how our internal teams would interact with them.

We agreed that Kennedy would receive all structural hardware first. Selected items could be sent to labs afterward with joint approval of our KSC team and the CAIB. Memory devices, tapes, and films were to go directly to the data team in Houston for analysis.

Representatives from our team had visited fifteen sites in California and found no credible shuttle debris there. A resident of the Yosemite/Tahoe area had heard twin sonic booms and found what he thought was a piece of tile from the shuttle. (It turned out to be a Styrofoam cooler.) However, a sheriff's department found what appeared to be a piece of tile with black specks, which needed closer examination. We sent another team to the Phoenix area to investigate possible debris sightings.[27]

Logistics for all the KSC people arriving at Barksdale and Lufkin was proving difficult. United Space Alliance's Rikki Ojeda and Linda Moynihan were using their personal credit cards to hold rooms for KSC staff coming into town. Both had run up several thousand dollars of charges. I contacted Kennedy and requested more administrative assistance.

—

Massive search efforts proceeded despite cold temperatures and a steady rain throughout the debris field. Limited visibility restricted some air search operations, but ground crews continued to work.[28]

Debris teams were asked to concentrate on locating reinforced carbon-carbon panels from *Columbia*'s left wing, material from the shuttle's outer mold line, supporting structure, and tile from the shuttle's belly. Analysis of the telemetry implied the presence of hot plasma inside *Columbia* prior to the accident, but it was still unclear how it had entered the shuttle's interior. Possible entry points included breaches in the left wing's leading edge, the left landing gear door, or from missing tiles on the shuttle's belly or underside of the left wing.

Despite the televised warnings about potential dangers, local residents were bringing in debris that they had collected. In the Lufkin Civic Center, astronaut Brent Jett called Ed Mango over to the astronauts' table to show him one item that someone had dropped off. Mango identified it by its distinctive shape as the door to one of *Columbia*'s star trackers—optical and electronic devices in the ship's nose that were part of the navigation system.

Another citizen brought in one item that he had found at his child's school. Mango instantly recognized that it was a pyrotechnic device. Not wanting to create a panic about a potentially armed and unstable explosive device in the middle of the command center, Mango nonetheless found himself backing away from the table, his face bright red. He thanked the citizen and sent him on his way. After he left, Mango called out, "I need a metal box, and I need it now!"

The collection centers continued to receive hazardous items from the field. When NASA started planning to ship the collected propellant tanks and pyros from Texas to Barksdale, regulations got in the way. Transporting potentially hazardous material across state lines was considered a violation of EPA regulations. Storing the material at Barksdale also exposed the air force to liability under the Resource Conservation and Recovery Act. After lengthy discussions and legal research, EPA eventually determined that it had authority to move hazardous debris from Texas to Louisiana, since both were officially declared emergency sites.[29]

By the evening's check-in call, Carswell Air Station near Fort Worth reported 198 pieces of confirmed debris in their hangar, including a piece of the leading edge of the left wing, its underlying carrier panel, as well as other material from that wing. They also reported that the Civil Air Patrol in the Dallas area was actively supporting airborne search efforts.

More than one thousand pieces of debris had been located in Sabine County, with about four hundred cataloged.[30] The collection center in Nacogdoches was still receiving forty to fifty calls per hour about debris sightings.

Lufkin reported that two recovered nitrogen tetroxide tanks had been completely decontaminated. Six EPA technicians were exposed to hazardous chemicals during the cleanup. The armed pyrotechnic

T-handle that had been turned in to Ed Mango was now secure in a munitions case.

Recovered pieces of the outboard elevon from *Columbia*'s left wing came in to the Palestine hangar. Other reported finds included a harness, a boot, strips of clothing, and an eight-foot section that might have come from the shuttle's left wing, with dozens of thermal tiles and heat sensors still attached.[31]

Several items from the avionics equipment area of the crew compartment were discovered in San Augustine County on Thursday. NASA used this information to request that a special team search the area northwest of Bronson on Friday for a classified box that had been in the same section of the shuttle's avionics bay.[32]

—

Searchers near Palestine found a videotape cassette on Thursday—one of many that at least partially survived *Columbia*'s breakup. The tapes were collected and sent to Houston. Several days later, astronaut Ron Garan took them in his T-38 jet to the NTSB headquarters in Washington, DC, to see if any information could be recovered from the recordings.

Garan and an NTSB technician spent hours playing back the tapes, which in many cases were charred or damaged. Most were either blank or contained data from the scientific experiments in *Columbia*'s Spacehab module. As midnight approached, Garan phoned Houston to report that he had seen nothing relevant to the accident investigation. After the call, he discovered that there was one more tape to check. He began playing it, and within a few seconds, he froze.

It was the cockpit video of *Columbia*'s reentry.

"The hair stood up on the back of my neck," Garan said, "because we didn't know how long the video was going to last or what it was going to show." To his relief, the tape ended several minutes before the first sign of trouble.

NASA publicly released the video on February 28. The tape showed *Columbia*'s crew being happy, acting professionally, and enjoying the ride. They were passing the video camera around, smiling at one another, and remarking on the sight of the glowing plasma surrounding the orbiter. They were obviously unaware that anything was wrong with their ship.[33]

—

At nine in the morning, on Friday, February 7, Kennedy Space Center conducted its memorial service. Bob Crippen, *Columbia*'s first pilot, delivered a moving eulogy for his beloved *Columbia* and her crew:

> It is fitting that we are gathered here on the shuttle runway for this event. As Sean [O'Keefe] said, it was here last Saturday that family and friends waited anxiously to celebrate with their crew their successful mission and safe return to earth. It never happened.
>
> I'm sure that *Columbia*, which had traveled millions of miles, and made that fiery reentry twenty-seven times before, struggled mightily in those last few moments to bring her crew home safely once again. She wasn't successful.
>
> *Columbia* was hardly a thing of beauty, except to those of us who loved and cared for her. She was often bad-mouthed for being a little heavy in the rear end, but many of us can relate to that. Many said she was old and past her prime. Still, she had only lived barely a quarter of her design life. In years, she was only twenty-two. *Columbia* had a great many missions ahead of her. She, along with the crew, had her life snuffed out in her prime.
>
> There is heavy grief in our hearts, which will diminish with time, but it will never go away, and we will never forget.
>
> Hail Rick, Willie, KC, Mike, Laurel, Dave, and Ilan.
>
> Hail *Columbia*.[34]

Immediately after the service, Sean O'Keefe and Bill Readdy went to the hangar at the south end of the runway. Shuttle Test Director Steve Altemus, who was setting up the hangar and who would manage the daily activities of the reconstruction effort, showed them the facility and walked them through how NASA would examine and reconstruct *Columbia*'s debris.

—

Search teams continued to explore areas in Toledo Bend Reservoir where debris might be underwater. They used side-scan sonar and aerial overflights to identify several sites for further exploration.[35] Aerial spotters occasionally thought they saw things below the surface. Whatever the objects were, they were gone by the time the search teams reached the area.[36]

At five o'clock that day, the amnesty period for citizens to turn in space shuttle debris in their possession expired. About twenty people had taken advantage of the amnesty to return material to NASA. Anyone subsequently found with illegally removed shuttle material would be subject to prosecution.

Late in the afternoon, Pat Adkins investigated a call from someone who thought that they had located part of *Columbia*'s side hatch. He met with Department of Corrections guards who were mounted on horseback. They had been searching the Indian Mounds Wilderness Area, near the reported locations of a car-sized object that crashed into Toledo Bend Reservoir and another large object that landed somewhere in the woods at the same time. The ground was muddy and waterlogged, and the area was thick with downed trees.

Adkins put on his backpack and said, "Let's get started."

The guards looked at him and said, "You don't understand. It's a mile and a half back in the woods. You'll never make it in and out before dark. You're gonna have to ride."

Adkins was uncomfortable about riding horseback. The guards gave Adkins the gentlest horse they had—George, a gelding, more than twenty years old. Adkins mounted the horse and followed the guards back into the woods. George was apparently tired after a long day's work. He stopped several times along the way to suck up muddy water from puddles. Then, realizing he had fallen behind the other horses, he would trot to catch up. Adkins had absolutely no control over him.

They came to a ravine, which the first three men jumped over on their horses. However, they cautioned Adkins, "You might want to get off of him." It was too late. Adkins held on to the front and back of the saddle and managed to stay on when George jumped the ravine. The guards laughed uproariously. One of them said, "That was worth the trip today!"

They found the object, and Adkins dismounted to examine it. After he knocked some of the mud off of it, he realized that it was not part of the hatch after all—just part of the propulsion system. However, as he stared at it, he began to imagine it as a hatch and porthole. His thoughts turned toward *Columbia*'s astronauts inside the crew compartment, and he broke down in tears.

After collecting his emotions, Adkins bagged the item and put it in his backpack. He rode George back to the road. Every time George trotted to catch up with the other horses, the heavy metal object in Adkins's backpack flopped up and hit him in the back of the head.

—

On Saturday, February 8, NASA's teams in Barksdale, Lufkin, and the Carswell Naval Air Station briefly stood down to attend memorial services for *Columbia*'s crew.

Former astronaut Jim Halsell led the service at Barksdale. He was the Shuttle Program manager at KSC, and he had asked me early on what he could do to help. I requested that he arrange the memorial service for his fallen comrades, which he did with dignity and strength. Retired Admiral William Pickavance, now USA's deputy director of Florida operations, brought with him from KSC the banner that decorated the launchpad gate before *Columbia*'s final flight. Having the Go Columbia! banner hanging on the wall in Barksdale was both a painful and poignant reminder for the workers who had cared for her at KSC and who now had to sift through her debris.[37]

From Houston, Ron Dittemore issued orders for the investigation teams to take Sunday off and rest. We had endured a week filled with long days of emotionally and physically demanding work. Even though we were all exhausted and conditions were miserable, no one wanted to stand down. In Carswell, at the northwestern end of the search area, freezing temperatures, ice, and snow tormented people who were bone tired from having searched fourteen hours per day for the past seven days. And yet, the search teams still went out after the memorial service—just as they had every day for the past week—to look for and recover more debris from the shuttle.[38]

Meanwhile, imagery analysis of *Columbia* in flight was yielding tantalizing results. NASA was able to piece together nearly continuous video coverage of *Columbia* in flight from the coast of California until the shuttle broke up over Dallas, thanks to dozens of amateur videographers who provided their cameras and tapes to NASA. Almost from the moment the ship was visible over California, her plasma trail unexpectedly brightened on occasion—apparently as tiles peeled off from the ship. NASA was evaluating the remote possibility *Columbia* had collided with some sort of debris or a micrometeorite as it crossed California.[39] We analyzed the weather radar along *Columbia*'s flight path as she flew across the United States and determined that there was no correlation between the "debris shedding" events and the local weather.

Observers at the Starfire Optical Range at Kirtland Air Force Base near Albuquerque, New Mexico, had obtained a puzzling image as *Columbia* flew past, about three minutes before the shuttle disintegrated. This disturbing photo showed apparent irregularities in the flow across the leading edge of *Columbia*'s left wing and something—possibly debris or vaporized metal—trailing out behind the left wing.

The pilot of an Apache helicopter, who was returning to Fort Hood from a night training mission when *Columbia* broke up, recorded a particularly important video. Seeing unusual streaks in the sky ahead of him, the pilot trained his targeting cameras on the smoke trails. Realizing later that he had witnessed *Columbia*'s disintegration, he personally drove the tape to Barksdale and played it for Dave Whittle and our leadership team. The tape itself was classified, but he allowed us to record portions of the video showing the breakup. With the detailed knowledge of the helicopter's position at the time of the accident and the altitude and azimuth data on the video, we gained crucial information about the shuttle's trajectory when it broke up, the dynamics of the disintegration, and the path of some of the major components after the shuttle came apart.

In another stunning development, we learned the "Flight Day 2" object detected by the Air Force was *real*. Something that was about the size of a laptop computer—with the radar characteristics of a piece of reinforced carbon-carbon—had drifted away from the shuttle on the second day of the mission. It added to the mystery of what had happened to *Columbia*.[40]

Material arriving at Barksdale from Texas was being prepared for imminent shipment to Kennedy. There were already 561 pieces of debris in Barksdale's hangar, with many more on the way soon. Of the twelve thousand pieces of debris collected in the field so far, none provided obvious evidence of why the shuttle broke up.[41] The reconstruction team would have to perform more exacting analyses to see what clues the debris could provide.

—

At the end of the first week of operations, the debris search area included sixty-one Texas counties, covering nearly thirty-three thousand square miles, and affecting more than seven million residents. NASA had three search teams at work in California, one in Arizona, and one in New Mexico.[42] We had dubious reports of debris being found in twenty-six other states and Jamaica, Canada, and Grand Bahama. They seemed implausible, but they still needed to be investigated.[43] Material might be in some of the reservoirs in Texas. The US Navy's Sea Systems Command volunteered their services for search operations in the major bodies of water along *Columbia*'s flight path.[44]

Miraculously, no injuries had been reported from the shuttle's breakup. About 130 people had gone to hospitals in the impact areas because of concerns about health issues. But nobody was injured by the debris, and no one had been admitted to the hospital for treatment. Only minor property damage to a few structures was reported. Had the shuttle broken up only a minute or two earlier, its debris would have rained down over Dallas, and the situation might have been very different.[45]

—

Citizens reported large pieces of debris in three ponds in Palestine on Sunday. NASA promised that divers would be sent to investigate as soon as the human remains operations were complete.

Local wireless companies had by now set up temporary cell phone towers at each of the debris collection centers. For the first time, the debris teams had reliable cell phone coverage—at least near the storage areas. United Space Alliance began issuing cell phones to everyone deployed to Texas for the recovery. This was an unusual practice in 2003, when companies typically only provided cell phones to senior executives.[46]

FEMA's Scott Wells publicly thanked volunteer agencies that had been providing coffee, meals, snacks, and morale-sustaining support that warmed the spirits of personnel in the field and recovery offices. Recognizing the tremendous progress made during the first week to bring the human remains and debris recovery operations under control, FEMA's Deputy Director, Michael Brown, acknowledged interagency cooperation as the key to success of the recovery efforts to date.[47]

—

One of the crew remains search teams had found a cockpit window frame from *Columbia* deep in the Sabine County National Forest on Saturday, but they were unable to retrieve the heavy item. Gerry Schumann took the six members of his debris response team out to the woods to retrieve the frame on Sunday. After parking on the main road as close as they could get to the object's GPS coordinates, his team had to trudge more than a mile into the woods to find the piece. The team spent nearly the entire day carrying the metal frame out of the forest—taking turns lifting it by its corners, setting it down when people got tired, and maneuvering it around briar patches. Eventually, the team wrestled it back to their truck.

Schumann was elated that his team successfully retrieved the item after such an ordeal. After delivering the window to the Hemphill collection center, he bragged about the accomplishment.

Don Eddings, Marsha Cooper, and several other US Forest Service people on hand at the command post seemed unimpressed. Eddings asked, "Was that you guys that brought that window in?"

Schumann said, "Yes! Great find. It took us all day to get that thing out."

Eddings said, "All you smart-asses had to do was ask us. There's a fire road near there. We could have drove you guys within fifty yards of it."

Schumann initially felt both angry and deflated, like he had been made to look foolish and that he had wasted so much effort. When he calmed down, he realized this was a wake-up call. The NASA staff's vast technical expertise was not always useful here. It was no substitute for the on-the-ground knowledge that the local residents had of their county, its forests, and its back roads. On such matters, he was completely ignorant.

There was no shame in asking for help.

From that time forward, Schumann did not hesitate to ask the US Forest Service personnel for assistance whenever he needed to get to a location deep in the woods. He and Eddings formed a fast friendship that persists to this day.[48]

—

The data team from Johnson and our debris team from Kennedy continued to clarify our roles and responsibilities regarding material recovered in the field. JSC agreed that after they had checked recording devices to see if they contained any data, the items would be sent onward to Kennedy for physical analysis. Crew photos and other audiovisual records would be considered unrelated to the reconstruction, and sent to the Flight Crew Operations directorate in Houston. Any recovered personal items, such as jewelry, were to be deposited with the FBI for secure storage, sent directly to the reconstruction hangar at Kennedy for processing, and then returned expeditiously to the crew's families. Personal flight equipment (seats, buckles, pieces of suits, and so forth) were to go directly to KSC for processing, but would be retained by NASA for the investigation.

On February 10, we drafted a memo of understanding such that the *Columbia* Accident Investigation Board had to concur before any piece of *Columbia* was subjected to testing, particularly destructive testing. This ensured evidence was preserved for the accident investigation—which, Administrator O'Keefe and Admiral Gehman told the press, they expected to complete in about sixty days.

Every collection center had already sent—or was preparing to send—shipments of recovered material to Barksdale. One truck from Palestine carried 276 items. Another truck was en route from Nacogdoches with ninety-three items. More than eleven hundred were still in the hangar there, awaiting transport. A truck from San Augustine was bearing forty boxes of material, which included a hydrazine tank that had not been decontaminated, a laptop disk drive, the faceplate from one of the shuttle's computers, a seat frame, one of the window frames, and a large piece of an external tank disconnect door. A truck was scheduled to leave Hemphill on Tuesday with more than two hundred items, including the frames to the six forward cockpit windows and a maneuvering thruster. Seventy-six packages of material, with nearly three hundred fifty items, were on their way from Jasper. One of these items was cockpit control panel A12, which came from the aft flight deck.

Although the Jasper collection center was full, very little additional material was being recovered nearby. We directed the Jasper site be closed once the last items in storage there were shipped to Barksdale.

—

The debris team in Lufkin brainstormed ways to locate debris from *Columbia*'s left wing in the area around Carswell, near the western end of the debris field, since most of the ground search forces were committed to looking for *Columbia*'s crew in Hemphill and San Augustine. Blimps, hang gliders, and powered parachutes were among the options discussed. The Civil Air Patrol had already made nine possible sightings from the air.

Teams from Alliant Aviation flew power-chutes (basically go-karts suspended from parachutes) over the debris field. On February 10, one group located a large piece of one of *Columbia*'s main landing gear doors on the ground. They landed nearby and directed a ground search crew to the location near Nacogdoches.

That night, they reported the find to Jeff Williams at the Forest Resources Institute, who was attempting to use satellite imagery to assist in the search-and-recovery effort. Williams discovered that he was able to see the gear door through the treetops on a recent classified IKONOS satellite image of the area. Williams was subsequently able to use this knowledge and his image interpretation skills to guide searchers to particular areas, based on what appeared to be broken tree crowns and branches that he could discern in the satellite images.[49]

Water search crews from the coast guard, FBI, and local police forces were still at work in Toledo Bend Reservoir and the Sam Rayburn Reservoir.[50] Following up on an eyewitness report, an FBI dive team retrieved one of the shuttle's landing gear brake assemblies from the water on the Louisiana shore of Toledo Bend.[51] Astronauts had been guiding the dive teams, directing the search assets, and investigating reported sightings. The astronauts realized they were out of their element in managing such a huge effort and needed expert help to run the water search operation. After touring the Toledo Bend Reservoir on the previous day, the navy formally offered to lead the water search efforts.[52]

Most of the possible debris sightings in California, Nevada, and Arizona had been investigated and closed out. No debris had yet been found west of Texas, even though amateur videos clearly showed as many as a dozen instances of pieces coming off from the shuttle between the California coastline and when it broke up near Dallas.[53]

Questions began to arise on the appropriate level of effort to be spent recovering debris. Once the critical pieces of wreckage had been collected—and many still remained somewhere out there—how much more of *Columbia* needed to be recovered?

—

One of our priority pieces from *Columbia*'s underside had arrived at Barksdale—the "sawtooth doubler." This two-foot by two-foot plate—roughly in the shape of the orbiter itself—had been bonded underneath the orbiter and then covered with tiles. Because the shuttle's skin was uneven where the doubler was mounted, the tiles covering that part had to be thinner than the surrounding tiles so that the exterior surface of the tiles was smooth. Thinner tiles might not stand up well to the heat of reentry, so it was important to assess the condition of this piece. It was one of the key items in the fault tree of possible failure modes.

Technicians and engineers at Barksdale saw the doubler was badly scorched and melted around the edges. Entirely by coincidence, one of the people processing the piece was the man who had installed it on *Columbia* in the first place.

It devastated him. Sobbing uncontrollably, this member of our KSC team held the piece in his hands and showed it to me. He was convinced that he was responsible for the loss of *Columbia* and its crew. Frank Travassos, a main propulsion system expert from KSC, and I took him outside. We tried to console him and reassure him that the accident was not his fault.

—

The next day, February 11, another key item on the accident fault tree had been recovered—a pyrotechnic device from the left landing gear. The pyro had fired, which would have occurred as part of the normal sequence to lower the landing gear. What we didn't know was whether the pyro had fired before the orbiter broke up or afterward. One of the last signals received from *Columbia* indicated that the left landing gear was down and locked. We were almost positive that this was a spurious reading because of heat damage to wiring inside the left wing, but we needed to examine this physical evidence as confirmation.

Meanwhile, Ralph Roe from the Orbiter Project Office was already thinking about ways to test the stiffness of the wing's leading edge reinforced carbon-carbon panels, another possible point of failure. He formed a team to investigate what kinds of changes to *Columbia*'s exterior surfaces could create the subtle drag that appeared to be causing some unusual movement of the ship's control surfaces before the orbiter broke up. Data showed *Columbia*'s steering thrusters were firing continuously—apparently trying to counteract drag on the left wing—just before the orbiter broke up. Roe was also interested in determining what kind of environmental changes could cause a breach of the seal on the left landing gear door.

My boss at KSC suggested Ed Mango be formally freed up from reporting to Barksdale so that he could concentrate on his work in the field. In effect, this was already happening. Mango had quickly become a key member of the leadership team at Lufkin, and I easily agreed.

In another management move that day, I learned I was going home to KSC in the next few days. Shuttle Program manager Ron Dittemore, at Barksdale with Linda Ham for a tour of the facility, asked me to return to KSC as soon as possible to head up the reconstruction effort. I was excited to be able to take on a new challenge. Being at the center of the action and solving problems energized me. Barksdale was definitely on the periphery now, and things there had become relatively routine. KSC's shuttle ground operations manager Dean Schaaf would be taking over for me. I knew I would be leaving things in good hands.

Later that day, I walked by the hangar where workers were carefully crating the shuttle material collected so far. They loaded the boxes onto two tractor trailers for transport to Kennedy Space Center. I watched the trucks pull away.

As I gazed at the wooden crates, a profoundly sad thought suddenly struck me: Columbia *is going home in a coffin.*

It seemed such a strange coincidence that the first pieces of *Columbia* herself would start going home to Kennedy Space Center on the same day that the last of *Columbia*'s crew members also left Barksdale for their eventual return home.

And now I would also be returning home—to receive *Columbia*'s wreckage, and try to determine the cause of the accident from it.

## Chapter 9

# WALKERS, DIVERS, AND SPOTTERS

With the news from Hemphill that the final *Columbia* astronaut had been recovered on February 11, the search effort now lacked the urgency of finding the crew. Meanwhile, the rest of the country was on war footing, preparing to invade Iraq. Rumors abounded that Homeland Security would raise the alert level in the coming days or hours. The National Guard was going to be recalled from the search effort immediately, taking with them their hundreds of searchers and their Blackhawk helicopters. The remaining makeshift search teams had done their job remarkably well despite the cold, rain, and rough conditions.

But critical pieces of the shuttle still remained missing, without which NASA might never know for certain what caused the *Columbia* accident.

On the evening of February 11, the leadership team in Lufkin discussed their options for continuing the search for *Columbia*'s widely scattered debris within the ten-mile-wide, two-hundred-fifty-mile-long debris path. NASA's Dave King worried it was too inefficient and dangerous to continue to put volunteers and law enforcement officers in harm's way searching a huge area in an undertaking that was likely to take several months.

The experiences of the previous eleven days exposed logistical weaknesses in the search effort. It would be unreasonable to expect the local incident managers to continue to run things for the next several months. Many of them had day jobs, unrelated to their work in incident management, to which they needed to return. It was impossible to find and train the huge number of local volunteers and other searchers needed for a sustained effort. Finally, the small towns of East Texas could not be expected to continue to endure the hardship of supporting hundreds or thousands of searchers for months.

Astronauts Dom Gorie and Jerry Ross were still figuring how to wrap their arms around the huge scope of the debris recovery effort. As they sifted through the problem, it became clear that they needed to organize operations in much the same way that Jim Wetherbee had organized the search for the crew—with officers appointed to the lead ground, water, and air search components. The US Navy had just agreed to take on water search operations, but who could provide the resources for the air and ground operations?

From the very outset of the crisis, local leaders who had large-scale incident management experience saw that this was going to be a long-term operation, based on the size of the debris field, the terrain, and the thousands of pieces of the shuttle that made it to the ground. Several of them, including Olen Bean, Marcus Beard, and Mark Stanford, suggested to FEMA that US Forest Service incident management teams (IMTs) could run the search efficiently utilizing wildland fire crews.

Their suggestions did not gain an immediate foothold, at least not until late in the second week of February.[1] It seemed illogical to NASA and FEMA to bring in wildland firefighting crews for a search-and-recovery operation. Perhaps it was interagency unfamiliarity with what IMTs and fire crews could actually do and the types of incidents in which they had already performed admirably.

The National Wildfire Coordinating Group, with membership from nine federal agencies, used IMTs to manage responses to large-scale fires and other major natural and man-made disasters. But those IMTs were not just professional firefighters. Their support structure included leaders and members from federal, state, and local agencies who dealt with an amazing array of "all-hazard" incidents. In addition to containing large-scale wildfires, they provided assistance in such diverse situations as search and rescue following the World Trade Center attacks on September 11, 2001, containing the Exotic Newcastle Disease (avian flu) outbreak in Nevada in 2003, and disaster recovery in areas devastated by hurricanes and floods.

When it became clear that the National Guard was being pulled from the *Columbia* search operation in the ramp-up to the Iraq War, Mark Stanford of the Texas Forest Service and Marc Rounsaville of the US Forest Service once again proposed that NASA and FEMA run the debris searches with IMTs managing US Forest Service fire crews.[2] This time, the circumstances were such that Stanford and Rounsaville's proposal looked very attractive. Dave King, Scott Wells, and Dave Whittle began to consider the idea seriously.

Stanford and Rounsaville pointed out the US Forest Service IMTs deployed as completely self-contained units and could be on-site within a matter of days. They brought their own tents, equipment, clothing, food, portable toilets, shower facilities, transportation, cooks, accountants, command organization—everything they needed to set up operations wherever their presence was required. IMTs normally utilized elite Type 1 "hotshot" crews and Type 2 fire crews to respond to complex incidents.[3] Each crew was staffed by twenty able-bodied, disciplined, and motivated men and women who were already trained in the practice of grid searching in remote areas. Hundreds of such crews existed across the country, with thousands of potential searchers. They would only need orientation in what to look for and how to handle what they found in the field.

Wells, Gorie, and Ross were intrigued. They asked Stanford how many people the US Forest Service could bring in. He said that the wildland fire service could bring in one thousand firefighters within a few days. Gorie and Ross immediately asked, "How about a thousand more?" Stanford said it could be done, but it might take another few days. When Gorie and Ross pushed for yet another thousand firefighters, Stanford pointed out that since February was not fire season, many personnel in the seasonal fire crews were currently off duty.

They would have to be mobilized, but it would only take a few days. The good news was that they were unlikely to be pulled away from the search to fight wildland fires elsewhere at this time of year.

Gorie was impressed. He said, "It was miraculous. I had no idea anything like that could ever be generated just for this effort."

Then came the question as to how to organize the search to maximize the probability of finding debris of a certain size, while keeping the resources and time required at a reasonable level. NASA's database indicated about 75 percent of the ten thousand pieces of material found so far was within two miles of the centerline of the debris path. Gorie asked that the Texas Forest Service run calculations for finding an object six inches square, which would be typical for a piece of tile on the shuttle's underside. Their analysis showed that a search line of people spaced five to ten feet apart—basically at just about fingertip distance—could find about 75 percent of such pieces on the ground if they searched two miles on either side of the centerline. To increase the likelihood of finding the occasional outliers, helicopters could survey the area on five miles on either side of the centerline. This ten-mile-wide path encompassed 95 percent of the debris already in the database.[4] Finally, water teams would search within two miles of the centerline in Toledo Bend Reservoir.

NASA liked the approach. On February 12, NASA and FEMA gave the Texas Forest Service the "Go" to contract for one hundred twenty-person fire crews, with their associated overhead and support personnel, and twenty-four helicopters from the US Forest Service. The US Forest Service agreed to have the helicopters begin air searches by February 14, with all ground crews searching in the field within two weeks.[5] More than half of the Texas Forest Service's employees had already been mobilized into leadership action for the *Columbia* search-and-recovery efforts.[6]

KSC had no shortage of people eager to help out in the search for *Columbia*. In fact, the hardest part of staffing the operation from NASA's perspective was turning down KSC personnel who wanted to help but who were not physically or emotionally up to the grueling conditions. "We didn't want to run the risk of hurting more people," said Dave King.

Assuming the US Forest Service's resources were in place by the last week of February, FEMA estimated that search operations of the entire debris field could be wrapped up by April 15. That date was important for several reasons. First, it was critical to try to beat the spring "green up," after which new foliage would render many of the wild areas nearly impassable and make it more difficult to spot debris on the ground. Second, NASA was under pressure to complete the investigation as soon as possible. Finally, the operation was expensive—the search effort would cost FEMA (and the US taxpayers) roughly $1 million per day.[7]

How could it possibly be worth spending so much money to recover wreckage from a destroyed spaceship?

Public safety was the overarching concern. Federal and state governments wanted all potentially hazardous shuttle debris collected and removed as quickly as possible.

Next, NASA and the CAIB needed to find physical evidence that proved conclusively how and why *Columbia* broke up. None of the thousands of pieces recovered so far appeared to be the smoking gun. Where had the breach in *Columbia*'s heatshield occurred that allowed hot plasma to penetrate the wing? Was there a hole in the wing's leading edge? Had the landing wheel well somehow been compromised? Or had tiles on the wing's underside burned through?

Was the *Columbia* accident a fluke, or did it point to a fatal flaw with hardware or procedures that could doom future flights? Unless we could prove with certainty what caused the accident, the rest of the shuttle fleet would remain grounded—perhaps permanently.

Without a shuttle fleet, the International Space Station could not be completed. Billions of dollars worth of space station modules were sitting in a processing facility at KSC, waiting to be launched. The United States would default on its commitments to its international partners, because the modules they had paid for and built could never get to the station without America's space shuttle. The shuttle was a critical asset, and it was in the country's interest to return the shuttle to flight.

FEMA, EPA, NASA, and the US and Texas Forest Services geared up for two months of an all-out push. By chance, the transition to the new search approach coincided with an unrelated need to move the joint command operations out of the Lufkin Civic Center. Lufkin wanted its prime community space back again, as a convention was scheduled in the coming week. Over the Valentine's Day weekend, the incident command team moved its operations to a bank building with vacant office space eight blocks south of the Civic Center.

NASA also decided to consolidate its command structure. Having two command centers—one at Lufkin and one at Barksdale—contributed to confusion and occasional contradictory responses to questions from personnel in the field. By February 28, NASA made Lufkin its command headquarters for the recovery effort. Barksdale would only be a staging area for shipments to Kennedy. Jerry Ross and Dom Gorie took turns leading the debris recovery effort for NASA.

Searching the designated *Columbia* debris corridor in Texas was now solely the responsibility of the interagency management teams directed by the Texas Forest Service. Volunteer search operations in Hemphill wound down, and the commanders there prepared to transition the searches to the professional teams. Many of the volunteers were disappointed that they could not continue to assist in the search.[8]

However, volunteers still aided the searches in areas outside of the main debris corridor. For example, many of the townspeople of Maypearl, Texas, turned out for a three-day community search effort that involved "four-wheelers, horses, and lots of free pizza."[9] Much farther west, volunteers searched unsuccessfully for *Columbia* debris along the California coastline beaches over the weekend of March 1.[10]

—

The Texas Forest Service arranged for the incident management teams to set up camp at fairgrounds or rodeo arenas in towns along the shuttle's ground track—Hemphill, Nacogdoches, Corsicana, and Palestine. These towns were close to the centerline of the debris field and were spaced roughly fifty miles apart. Each town's camps could accommodate eight hundred to one thousand searchers. At any given time, five IMTs would be in Texas, and the IMTs and their fire crews would rotate out every two to three weeks. Twenty-one IMTs eventually participated in the search-and-recovery operations.[11]

The IMTs and their fire crews entered and departed East Texas through a staging and coordination center at Longview. Longview oriented incoming search crews on the conditions they could expect on the ground in Texas. No motels were available, so searchers needed to bring tents and sleeping bags. Temperatures could range from the mid-twenties at night to the mid-seventies in the daytime. Heavy rains could occur at any time. Searches would be in terrain that varied from open pastures to swamps to extremely dense vegetation. Ticks and chiggers were common, although the cool temperatures were likely to slow down any snakes and alligators in the area. Finally, searchers might encounter hazardous materials or human remains.[12]

The Southwest Texas Debriefing Team, a part of the Critical Incident Stress Management network, spoke with fire crews when they first arrived in Longview. Searchers received information on how to monitor themselves and their colleagues for signs of physical and emotional stress. The briefing team also reminded searchers how grateful NASA and the nation were for their efforts.[13]

Crews came from almost every state in the United States. Much to the surprise of the NASA staff, most of the crews from the Midwest and Western United States consisted of Native American firefighters, representing nearly every Native American Tribe and Nation in those states. In many areas or reservations, it can be hard for young adults to find jobs. Working on a fire crew both provides good temporary income for able-bodied persons and also helps protect tribal lands during fire season. The presence of people representing so many different tribes provided a multitude of lessons in cultural diversity.

By February 23, 4,560 wildland fire personnel were on-site in Texas, which included 169 twenty-person crews and 1,272 overhead support positions.[14]

Jerry Ross recalled seeing the fairground area at Nacogdoches before the searchers arrived, when it was just open field and concrete slabs. He returned two days later to find ". . . hundreds of people established there with all their equipment, their tents, their cooking facilities, their showers, their bathrooms—everything. *Pow!* It was there! You make a phone call and things start happening immediately. It was pretty amazing."

Greg Cohrs said that with the arrival of the fire crews, the Hemphill Camp at the rodeo arena became the largest "city" in Sabine County.

—

Search coordination evolved into a sophisticated operation two weeks into the recovery of *Columbia*'s debris. The Texas Forest Service produced a search plan using two-mile search grids along the centerline from just west of Fort Worth in Texas to Fort Polk, Louisiana. The grids were assigned alphanumeric codes, and search teams throughout the debris corridor used the same coordinated system for the remainder of the debris recovery operation. To be certain they missed nothing, much of the already-searched part of the corridor—including the entire San Augustine and Sabine County area—was covered again, as if it had never been searched.[15]

Barksdale, Lufkin, and the reconstruction hangar at Kennedy exchanged detailed information on what was being retrieved and where it was coming from. As more debris and data came in, it became easier for us to identify the areas in the field where particular items of interest might be found. For example, most of the crew module items were being found in San Augustine and Sabine Counties. If we identified in the reconstruction hangar a circuit board from a component of one of *Columbia*'s avionics boxes, teams searching for a sensitive computer that had been near that component on board *Columbia* could be targeted the next day to search in the area where the board was found.

NASA's "ground boss" in Lufkin coordinated with the rest of the leaders to review what the previous day's items of interest were and which areas to focus on next.[16] The Texas Forest Service printed maps using recent satellite photos of the targeted search areas and distributed them to the appropriate command centers every evening.

Every morning, the IMT commanders at each of the search centers reviewed the maps and instructions with their strike team leaders, who typically supervised five twenty-person crews,[17] and assigned each team's search area for the day. Then the crews would disperse to their assigned areas and begin searching.

The influx of technical support from KSC continued as the next phase of operations ramped up. These were primarily United Space Alliance technicians, although some NASA employees also participated in the search teams. Regardless of whether this person was a civil servant or a contractor, search crews referred to their assigned KSC representative as their "NASA."

One NASA and one EPA contractor accompanied each fire crew on a day's search. More than two hundred of our workers from KSC, and about the same number of EPA contractors, were on the ground across the Texas debris field every day.

Because Texas was still under a federal disaster declaration, the search teams technically had the authority to search private property without a landowner's permission. An FBI representative accompanied each search team that crossed private property in case official authority needed to be exercised. Most of the time, landowners were willing to let the searchers onto their property. On occasion, an armed farmer or rancher refused access. Rather than try to press the issue, the search team leaders usually backed off and asked the landowner to check the property himself.[18] As one person pointed out, "Texans will give you the shirt off their back. But if you try to take it from them, they'll fight you for it."

Livestock ranchers, as well as poultry and dairy operators, were understandably concerned about large groups of people traversing their land and frightening their animals. Chickens might not lay eggs for several days if they were disturbed. Poultry could also suffer heart failure, suffocate, or be exposed to viruses. Spooked cows and horses could run into barbed wire fences and injure themselves. NASA and FEMA legal teams in Lufkin eventually processed 153 claims for damages resulting from the search operations.[19]

Territorial bulls were also problematic. NASA's Gerry Schumann and Debbie Awtonomow, who was managing the Hemphill collection site, went with a team one day in March to retrieve debris reported by one rancher, which turned out to be battery packs from experiments in the Spacehab module in *Columbia*'s payload bay. When Schumann and Awtonomow tried to return to their vehicle, a bull blocked their way. It seemed to be infatuated with Awtonomow and would not leave her alone. Some of the team distracted the bull long enough for her to get back in the vehicle. Then the bull stood in front of the car and put its horns against the grille, refusing to let the team leave. The rancher had to lead the bull away.

René Arriëns, a member of our shuttle's closeout crew, now found himself in a very different world from working in the White Room at the launchpad. He spent his nights in a trailer at a fish camp on Toledo Bend. By day, he walked with one of the two search teams under his direction. "I thought I was fairly prepared for anything they needed," he said, "but it wasn't anything like I thought it was going to be." At times the terrain was swampy, requiring the search team to spend the day walking through water halfway up their shins. Arriëns counted fourteen different types of thorny bushes, all of which tore up his clothing so badly that he had to buy a new shirt or pants every day or two.

It rained all but three days during the time that Arriëns was in Sabine County. Deep mud seemed to be everywhere, and workers could not avoid tracking it wherever they walked. People traveled in groups so that someone could help push cars out of the muck if necessary. Mud would splatter up into the engine compartment of a car, and if allowed to dry, it could become as hard as concrete, ruining the vehicle. Felix Holmes from the US Forest Service occasionally had to use his bulldozer's winch to dislodge search vehicles stuck in ditches and creek crossings.

On sunny days, snakes seemed to be basking everywhere. People often came close to stepping on them, causing searchers to jump and run. Arriëns encountered the biggest coral snake he had ever seen. He even saw snakes up inside bushes.

The woods held other hazards for those who did not know what to watch out for. On one of the few clear days during his deployment, Arriëns and his crew were working in an area where timber cutting or a beetle kill of trees had occurred. These kinds of forest openings often contained dead snags, root-sprung trees, or trees weakened by insects or disease. Workers in such areas had to pay close attention to "widow makers" that might break or be blown over in the wind or rain. Arriëns and two of his colleagues sat down on a large fallen tree to eat their lunch. They suddenly heard a loud *crack!* and a twenty-inch diameter tree fell at their feet. One foot farther to the side and it would have killed them instantly.

Arriëns only fell behind once during a search in a dense stand of young pine trees. He left his spot to examine what appeared to be parts of a TV system. However, he neglected to make a note of his position before he walked off, and he became disoriented. "The pine needles were right in your face," he said. "You could talk to someone one tree over from you, and you could not tell where it was coming from, because the undergrowth was so very dense." The team started yelling to him to help him find his way back. When he finally found them again, several of the men had placed their gloves on their shovel handles in such a way that the handles extended the middle fingers of their gloves. It was a sarcastic but good-natured reminder to follow safe woodland practices.

Arriëns brought his harmonica with him and would occasionally play when things got tedious on days where little was to be found in the woods. He thought that the others might not like it. However, it lightened the mood, and they often encouraged him to play.

Arriëns unexpectedly confronted his emotions on one day's search. He had been one of the last seven people to see *Columbia*'s crew on January 16, when he helped strap them into their seats in the orbiter before launch. Now, at a spot near a magnolia tree with a stream running beside it, he saw a cross and flowers marking where remains of one of *Columbia*'s crew members had come to Earth. "I just couldn't go any further," he said. "I stayed there for about forty minutes. That's where I put them to rest. And then I went on and did my job."

One of the Sabine County search teams suffered a scare on Sunday, March 2. KSC's Pat Adkins was walking with a search team in the Six Mile and Big Sandy Coves area near Toledo Bend Reservoir. Adkins heard over his portable radio a shouted order from the crew boss for everyone to freeze in place immediately. One of the searchers thought he had seen something buried in the ground. He knelt at the spot and poked at it with his walking stick. He and the people around him suddenly became nauseous, headachy, and short of breath. Everyone immediately feared that he had encountered a shuttle component contaminated with hazardous chemicals.

While the overcome men were being attended to, Adkins gathered the rest of the crew and gave them a safety reminder. "Nothing out here in these woods is as important as your safety. We're gonna find what we need to find, whether you take a chance or not. So don't take chances."

The EPA closed off the search area for a day so hazmat teams could clear it. They determined the searchers had been affected by naturally occurring swamp gas—methane with hydrogen sulfide and carbon dioxide—from rotting material in a stump hole. When the searcher poked his stick into the mat of vegetation over the hole, it released the trapped gas.[20]

Stump holes were familiar hazards to the local residents who had stepped in them many times during their years of walking in the woods. For people who were not used to the heavy rain and muddy conditions in the pine forests, their first encounters with the hidden traps could be scary. Jamie Sowell was with one Sioux strike team in the woods one rainy day when he heard screaming up ahead. One searcher had stepped into a stump hole with both feet, fallen in, and was now trapped up to his waist. His colleagues stood in a wide circle well back from him, believing he had been caught in a swamp of quicksand. Sowell said, "My guy walked over to him, grabbed him by the hand, and helped him out. Then he explained to everyone that it was a stump hole."

Weather extremes were difficult to deal with. The persistent cold rain and sleet of February began to alternate with hot days in the spring. Searchers from the Pacific Northwest and Alaska had difficulty coping with the stifling heat and humidity of Texas. Even though coordinators like Greg Cohrs monitored the forecasts and radar data as best they could, sometimes personnel were caught out in high winds or torrential rains. On one occasion, searchers were fortunate to be missed by a tornado, but were painfully pelted by hailstones.

The briars and thorns were as much a problem to the professional wildland fire crews as they were to the volunteers in the early days of the searches. Florida firefighter Jeremy Willoughby said, "I've been in the woods a lot, but I've never seen anything that thick. You had to use your body weight in some places to lie down in the bushes and make yourself a path. One guy next to me went underneath a branch and got a thorn right in his eye. It was horrible."

Pat Smith said that searchers covered with scratches and cuts would come into her bank in Hemphill. "I'd say, 'Oh, you're walking!' They would reply, 'Yeah, and those *wait-a-minute* bushes are killing us.' They called them that because while they were walking, someone would run into one and yell, 'Wait a minute! Wait a minute!'"

Through it all, NASA was deeply impressed at the keen eyes of the fire crews during the search. "They did not miss the slightest bit," Arriëns said of the Native Americans with whom he searched on his first day. "They'd come to me with pieces under a quarter of an inch." Debbie Awtonomow said, "They were phenomenal. One guy saw a puddle in the woods, and he knew from experience it didn't look right. They found a six-foot beam buried in the mud under that puddle."

—

The Forest Service initially staged its contracted helicopters at the airports in Lufkin and Palestine.[21] More than thirty helicopters were in the air at any one time over the search area, sweeping a three-mile path on each side of the ground search corridor. The Texas Forest Service flew fixed-wing aircraft over the area to coordinate air traffic. Astronauts Scott Kelly, John Herrington, Tim Kopra, Terry Virts, and Butch Wilmore took turns as NASA's coordinators for the air search.[22]

The helicopter contractors had their own fuel trucks, which helped to maximize the time the choppers were deployed every day. Pilots could arrange to have the fuel truck meet them at a pasture near the day's search area, rather than having to fly back to the airport to refuel. The spotters had time to grab a quick sandwich, and then it was time to get back into the air.

The aerial search crews usually consisted of a pilot and a helicopter manager from the Texas Forest Service in the front seats, with another Texas Forest Service or US Forest Service worker and one or two KSC engineers or technicians in the backseat to spot debris. If a clearing was nearby, the chopper could land to inspect or retrieve an item that looked particularly noteworthy. Otherwise, the crew noted the GPS location and sent a ground team to retrieve the piece of debris later. Helicopter spotters were also sometimes able to see items in areas that were inaccessible to ground searchers.[23]

Typical of the KSC staff flying as spotters was NASA test project engineer Mike Ciannilli, who flew with a search crew out of Palestine. He had never been in a helicopter before, but he desperately wanted to help in the search for his beloved *Columbia*. Several days of crippling airsickness at the beginning of his assignment almost forced him out of the role. However, someone suggested that he ask the pilot to fly with the helicopter's doors removed. The fresh, cold breeze helped reduce his nausea, and he was able to proceed in his role.

One of the challenges in spotting shuttle debris from the air was discerning it from the junk landowners left in remote corners of their property. Bathtubs turned up in strange places. On one flight, Ciannilli spotted what appeared to be a perfect piece of white tile in a swampy area that was miles from the nearest house. He asked the pilot to set down as close as possible to it, but the only safe area to land was several hundred yards away at the edge of the woods. Ciannilli slogged downstream through water and muck to reach the item, only to discover it was a Texas license plate.

Search teams needed to be aware of unofficial "no-fly" areas in the search area. On another of Ciannilli's flights, the pilot received a call from the airborne control plane, "Abandon your search! Abandon your search!"

Ciannilli asked, "Are we low on fuel?"

The pilot said, "It's not safe. We might get shot at. There's a meth lab in this area."

—

Dive teams from the FBI, EPA, coast guard, Houston police department, the Sabine River Authority, Jasper County sheriff's office, and other local authorities had been searching the Toledo Bend Reservoir since February 1, following up on reports of large objects hitting the water. Unfortunately, dense fog covered the lake at the time of the accident, and few people were able to judge reliably where anything actually impacted. Their reports were actually classified as "ear witness" testimony.[24]

The only thing recovered so far from Toledo Bend was a brake from one of *Columbia*'s main landing gear assemblies. An FBI team found it in shallow water a few feet west of the Louisiana shoreline. *Columbia*'s OEX recorder had not been found on the ground. NASA suspected it might have come down along the Texas side of the lake, since other items from the forward end of the shuttle were found onshore nearby.

The US Navy took over the water search and salvage operations in Toledo Bend and Lake Nacogdoches at about the same time the national Incident Management Teams took over the ground search.[25] The navy surveyed the areas of the reservoirs along the centerline of the debris path, and immediately saw the challenges they would be up against. Water temperatures were in the forties. Treetops, moss, and hydrilla just below the water's surface restricted the depth at which salvage vessels could tow sonar devices. Suspended matter in the water limited divers' visibility to just a few feet. Operators on the surface would need to use portable sonar devices to guide divers to areas of interest. Divers would have to wear special dry suits to protect them from potential exposure to toxic chemicals.

Searching for debris on the lake bottom was extraordinarily difficult, especially since the initial search area was over thirty-two square nautical miles of lake bed. When Toledo Bend's dam was constructed in 1966, engineers expected it would take three years for the reservoir to fill with water. That would have allowed enough time to log the forested land and reclaim items from buildings in the reservoir. However, two back-to-back major floods caused Toledo Bend to fill up in only three months.[26] There was not enough time to clear trees and remove equipment from the farms and logging operations, and all of that was now sitting at the bottom of the lake.

Consequently, the navy's sonar picked up returns from myriad large and small metal objects on the lake bed—railroad tracks, tractors, cars, storage tanks, fence lines, and metal roofs on storage buildings. Even flat-topped tree stumps created sonar returns that appeared similar to potential targets of interest.

High winds prevented search teams from working on the lakes on February 21, 26, and March 29, but teams were otherwise in the water every day. They examined anywhere from twenty to more than one hundred targets of interest every day.

By March 2, the navy had not located any material from the shuttle in either Toledo Bend or Lake Nacogdoches. The navy asked NASA for some actual space shuttle material upon which the search teams could test their equipment. NASA provided four pieces of shuttle skin that had been recovered on land. The dive team tied the pieces to buoys and dropped them in about twenty feet of relatively clear water near the Fin and Feather Resort. The search equipment could only reliably detect the largest piece, which measured about three feet by six feet across. The reservoir was on average sixty feet deep, and more than one hundred feet deep in places. This did not bode well for finding small debris from the shuttle with the current resources. A higher-resolution sonar system brought in a few days later also had difficulty detecting small objects.

Nonetheless, analysts became proficient at identifying man-made objects in sonar scans and targeting dive teams to them. The recovered objects—none of which were from the shuttle—were proudly displayed against the wall of the dive headquarters. These trophies included a drywall bucket, anchors, outboard motors, and a refrigerator door. The materials were all about the same size and shape as the debris that NASA was most eager to find.

On March 18, NASA asked the navy to look in one area of Toledo Bend for a camera that could have recorded pictures of the shuttle's external fuel tank. Someone said they had seen film or tapes along the edge of the reservoir, but they were lost when heavy rains raised the water level.[27] Several pieces from a camera were found on the western shore, raising hopes the camera body might be in the lake.

The navy scanned the new area several times during the search for the camera. Divers spent 176 diving hours grid searching an area of about five and one-half acres. This concentrated search produced only a single, thumbnail-sized piece of shuttle skin. While searchers were no doubt frustrated to have found nothing of consequence from *Columbia* in Toledo Bend, they demonstrated that searches with high-tech systems were not missing any significant pieces of shuttle debris.[28]

—

On February 18, a Sabine County resident who was four-wheeling near the bank of an inlet near Six Mile Bay on Toledo Bend spotted

*Columbia*'s nose landing gear partially buried in the mud. That afternoon, a pickup truck brought the nose gear to the collection center at the Sabine County rodeo arena.

As Pat Adkins cleaned off the piece with a hose and scrub brush, he reflected on how bizarre the situation was. Here he was, standing in the bed of a pickup truck, washing mud off of a once-pristine piece of the shuttle. Unlike some other items of debris that were almost unrecognizable, there was no mistaking what this was. Both tires were still attached to the strut. The bead of the tires was burned off, and they were deflated, but the tires were otherwise remarkably intact. The stroke arm for the steering actuator was missing, and Adkins could see that one portion of the surface of the strut had been exposed to the heat of reentry. After cleaning it off, Adkins and several other men transferred the landing gear to the back of a semitrailer, where it could be kept secure until the next shipment of debris from Hemphill to Barksdale.

The immediately recognizable piece of the shuttle evoked strong reactions in the NASA personnel. John Grunsfeld saw it on a visit to the Hemphill area. "He was obviously sobered by being in the presence of the item," Greg Cohrs said. "Then he told me that he had been on the last flight of *Columbia*."[29]

On February 21, NASA's Debbie Awtonomow came from KSC to manage Hemphill's collection center. Pat Adkins and Gerry Schumann familiarized her with the area and then offered to show her the latest recovered items. Adkins raised the door at the back of the trailer. Awtonomow looked in and immediately saw the landing gear. The sight of the piece of the once-proud shuttle, now horribly wrecked and embedded with grass and mud, proved too much for her. She walked over to the ramp beside the truck and vomited. She broke into tears and cried for nearly an hour.

"In the back of your mind," she later recalled, "you tell yourself that it's just a dream, that this is not really happening. But to see this the first thing—reality hit real quick. It was like someone took a two-by-four and smacked me upside the head."

Schumann and Adkins sat with her and comforted her. Adkins told her, "It's not going to get any better. It's good just to get it out of the way now. I understand."

When she had finished crying, Schumann consoled her, "We all went through it. We all had our time that we had to break down and get it out of us, and then go on and do the job. Are you ready to go to work now?"

The long hours, tough physical conditions, and the emotional challenges of the work took their toll on our NASA workforce and the local officials. People burned out quickly. Personnel usually stayed on site for several weeks and then went home to rest and recharge. On average, between forty and fifty people rotated in and out of the area from KSC every week.[30]

When Stephanie Stilson took over from Tom Hoffman at the Nacogdoches hangar in late February, she could tell at the moment she arrived that "Tom was definitely ready to go. You're working sixteen-plus hours per day, seven days a week. You can't do that for long periods of time and be as effective as you need to be."

—

The collection centers moved into high gear as crews retrieved the debris that was tagged but not picked up in the first two weeks following the accident. By February 18, over four thousand pieces of debris had already been sent to KSC, and ten thousand more items were on their way to Barksdale from the collection centers. By March 4, the number of pieces of debris found had more than doubled. The combined ground and air searches were gathering a great deal of material that could prove important to the investigation.[31]

Each ground search team usually located and retrieved between two and fifty pounds of debris every day.[32] Landowners were also still uncovering material on their properties. On February 13, a man plowing his field in Littlefield, Texas, north of Lubbock, found a small piece of tile. The "Littlefield Tile" turned out to be the westernmost piece of *Columbia* recovered—nearly three hundred miles farther west than any other item of debris—despite searches in every state between Texas and the Pacific Ocean. *Columbia* shed this piece of tile from its left wing about one minute before the vehicle completely disintegrated.

At the other end of the debris field, one of the space shuttle main engine turbopumps was pulled out of the mud at Fort Polk, Louisiana, on February 15. Another was found on March 30 and retrieved on April 1. Still traveling supersonically when they impacted the ground after the accident, the heavy pieces of machinery buried themselves fourteen feet deep. These dense engine components flew farther than any debris after the shuttle broke up, and their trajectory was unaffected by winds aloft. The path from the point where the shuttle broke up near Palestine to where the powerheads impacted near Fort Polk defined the initial centerline of the debris search effort.

In Sabine County, René Arriëns and Debbie Awtonomow responded one day to a report of a cassette tape stuck in the upper branches of a tree. Awtonomow first thought someone had thrown it from a passing car. Then it struck her: *How could it have ended up in the top of a tree unless it fell from the sky?* They retrieved the case and as much of the tape as they could. The case was scorched, but some of the tape appeared to be intact. It was a personal music cassette from one of *Columbia*'s crew.

Some of the experiments from *Columbia*'s Spacehab module survived reentry and made it to the ground. Pat Adkins responded to a call from a woman who found bags with aluminum cylinders hanging in the trees at a far corner of her property. On another occasion, Adkins wondered aloud to an EPA colleague whether any of the fish eggs on an experiment in the Spacehab module survived the accident. The EPA man blanched and said, "Oh my God, don't tell me that! I can just imagine a fish species that isn't native to Texas coming in and taking over."

Back at the collection centers in the evening, NASA personnel sorted through the material collected during the day's search. The initial triage consisted of segregating things into "definitely shuttle material" and "definitely not shuttle material" boxes. One piece of rusted steel in Sabine County appeared to be from a pickup truck. The man who first examined it tossed it in the "definitely not" box, because he did not believe any steel was on the shuttle. However, Arriëns knew the shuttle's landing gear strut was made mostly of steel, and he recognized the mechanism right away. It turned out to be a critical find—one of the first pieces of the shuttle to be exposed to plasma during reentry.[33]

Another piece that came in was a shard of something that looked like polished metal. No one could figure out what it was. Then the person holding it accidentally dropped it onto the pavement, and it broke. It was a fragment of glass coated on all sides with melted aluminum. Pieces with this kind of metal deposition would be crucial in reconstructing the sequence of events in *Columbia*'s breakup.[34]

Searchers near Powell, Texas, found one tile with puzzling orange deposits in jagged grooves on the surface of the tile. The markings did not appear to be reentry damage, but it was unclear whether the orange deposits were from foam that came off the shuttle's external tank. Whatever the cause, it supported the hypothesis that something collided with the shuttle in flight, although the idea of a collision during reentry was later debunked.

NASA's "fault tree" had been pruned by early March to the extent that there were now only ten different failure scenarios that might explain how heat had entered *Columbia*'s left wing. The leading edge of the left wing was the primary focus, but engineers still could not rule out a burn-through from the bottom of the wing. More hard evidence was needed.

—

On March 19, firefighter Jeremy Willoughby was searching with the "Florida 3" fire crew, one of several crews from Florida deployed to an area in San Augustine County, Texas, where material from the shuttle's crew module came down. Willoughby's crew was walking that morning through a pine stand on a gradual slope, when someone saw a metal box on the ground next to a small crater, where it had impacted and bounced. "It was just laying there like, 'Here I am!'" said Willoughby. The box was wrapped up and placed into the back of a pickup truck along with other items the crews found that morning.

Greg Cohrs and the FBI's Terry Lane were riding around to check on the progress of the day's searches in Sabine and San Augustine Counties. At the Magnolia Church staging area in San Augustine County, they looked in the back of a pickup truck holding material collected by the Florida crews. Cohrs immediately recognized what appeared to be an important piece of equipment, wrapped in a plastic bag. He thought it might be the shuttle's Orbiter Experiments (OEX) recorder, but he was not certain. He asked the group supervisor what he thought it was, hoping the supervisor would voice the same opinion. However, the box looked a little different from the photograph that NASA had used to alert searchers, and neither the group supervisor nor his on-site NASA counterparts seemed overly excited about it.

Convinced that NASA managers would want to see the component immediately—whatever it was—Cohrs and Lane took possession of the box and drove it directly to the collection center in Hemphill.

At the collection center, Cohrs and Lane watched as the staff removed the box from its bag and cleaned it off. Considering the box had survived hypersonic reentry from two hundred thousand feet in altitude and had been exposed to the elements for seven weeks, it appeared to be in remarkably good shape. Even its government property sticker was still intact. Cohrs noticed only a few damaged places, such as where connectors had broken off of the back of the case.

Greg Breznik, the NASA coordinator at the site, looked at the box and phoned Dave Whittle at Lufkin. He told Whittle the location and the number on the side of the box. Whittle said, "I want it on my desk!"

A few seconds later, Breznik told everyone, "Step away from the box! It's the OEX recorder!" With that confirmation, Cohrs and Lane returned to San Augustine to tell the Florida search crews the good news about what they had found that day.[35]

Upon hearing the news from Cohrs, Willoughby's crew agreed among themselves not to reveal the name of the person who found it. They wanted to remember the important moment as an accomplishment shared by the entire crew.[36] They found the box within an area that had previously been searched by the special team searching for a classified communications box on February 7. Cohrs did not realize this until he reviewed his notes several months later.[37]

NASA immediately drove the box to the Lufkin search headquarters. KSC's Jeff Angermeier was among the people there to witness the box's initial inspection. Many of the other avionics boxes from *Columbia* were burned, melted, or smashed almost beyond recognition. In contrast, the OEX recorder was almost pristine, except for the missing front panel and the torn connector holes in the back. There was virtually no evidence of heat damage. The amazing condition of such a critical item was another of the miraculous events surrounding the *Columbia* accident.

That evening, seven thousand miles away, *Operation Iraqi Freedom* began with the "shock and awe" bombardment of Baghdad. MSNBC's war coverage was briefly interrupted by news that NASA had recovered "*Columbia*'s black box" in the Hemphill area.

NASA shipped the OEX recorder to Imation Corporation on March 21 to inspect and clean the recorded tape inside the unit. Imation found that the tape had broken between the supply and take-up reels. However, the length of tape on the take-up reel implied that the recorder had started up as planned about fifteen minutes before *Columbia* began reentering the atmosphere. The tape was too damaged for any data to be retrieved at the point where the tape broke, which might have happened in the violence of the shuttle's disintegration. With luck, however, the recorded data might be complete up to the moment where the shuttle came apart and lost electrical power.

Imation shipped the box to KSC on March 25, where the tape was duplicated. The box and copies of the tape then went to JSC for analysis by the data team.[38]

The last few seconds of telemetry received in Mission Control on February 1 indicated *Columbia*'s crew likely knew their ship was in trouble in the final half minute before it broke apart. The data showed that *Columbia*'s steering thrusters were firing to compensate for drag on the left wing, the ship was rolling, and the triply-redundant hydraulic system was losing pressure. All of those conditions would have set off alarms inside the cockpit. If the OEX recorder's tape was readable, it would enable Houston's data team to determine the condition of hundreds of the orbiter's other systems throughout reentry, perhaps up until the moment the ship finally came apart.

The OEX recorder would not be able to tell NASA the story of what was happening to the parts of *Columbia* lacking instrumentation, though, such as the condition of the thermal protection system's tiles and wing leading edge panels. And Kennedy's debris team was still missing much of the physical evidence of what happened to the ship.

Coincident with the recovery of the OEX recorder, search operations reached the halfway mark on March 19, with 257,000 acres searched to date. More than 43,000 pounds of shuttle material had been recovered, representing 20 percent of the shuttle's weight.[39] By March 24, more than 10,270 firefighters and their support staff had worked the search operation.[40]

Solid progress was being made. And yet, the painstaking search of the debris field needed to continue for at least another month, until every square foot of the search corridor had been covered.

## Chapter 10

# THEIR MISSION BECAME OUR MISSION

The wildland fire crews proved to be remarkably efficient at recovering *Columbia*'s debris, working diligently and with great discipline. What may have started as a means for the crews to earn much-needed money in the off-season quickly turned into a profoundly meaningful experience. Working alongside the fire crews was also a life-changing experience for the NASA personnel, as they met men and women whose backgrounds were unlike any they had ever known.

Astronaut John Herrington was thrilled to work with the Native Americans who staffed many of the fire crews, particularly those from the Western United States. As an enrolled member of the Chickasaw Nation and the first Native American to fly in space, he felt a deep connection with these men and women.

Their attitude toward their search-and-recovery assignments fascinated him, particularly how each man or woman handled the pieces of debris they found during the searches. "These really rough, hard-core, no-nonsense, work-hard people would treat every piece they found with such reverence," he said. "It wasn't an inanimate object to them. Each item was very alive, very real. They understand that everything around us is a living, breathing being that we cooperate with. It made me appreciate my heritage, what these people sacrifice, and how special this experience was to them."

The crews from the various Tribes and Nations usually tended to keep to themselves. Under normal circumstances, they would primarily interact with their own people. Larger incidents occasionally brought crews from various Native groups together to fight a fire, and sometimes competitiveness and even old rivalries dating back centuries could rise to the surface. On the *Columbia* recovery, everyone—Blackfoot, Sioux, Creek, Cherokee, Choctaw, Apache, and others—generally put aside their differences and worked together.

One of the first incident commanders the Forest Service brought in was a man named George Custer, who administered the Nacogdoches camp. Understandably, many of the Native Americans wanted to have their picture taken with him.

The workers from KSC received an invaluable education in cultural diversity when they wandered around the tent cities of the various camps each evening. Many of the fire crews decorated their tents with tribal symbols or other items from home. Searchers drew pictures of *Columbia* and her crew or wrote poems telling the story of the crew's sacrifice and how it touched them. Each one of these "badges" illustrated profound reflections about the significance of the tragedy and its connection to a deeper spiritual meaning.

Tribal groups frequently sang songs in their Native languages at night. A hush would fall over the camp while a group chanted their song in their Native tongue. It was a side of America that most NASA workers had never experienced, and it heightened the utter uniqueness of the situation. The wails of the singing and the muffled drums in the darkness often caused KSC workers to reflect on just how far away they were from home.

NASA knew the importance of showing its appreciation to the searchers and helping them see the connection of their roles to the bigger mission. Mishap Investigation Team Chairman Dave Whittle said, "Those fire fighters were out there in horrible conditions—in water, with snakes, being chased by bulls, marching through rain, sleeping in tents in very cold weather. We wanted to make sure that people knew the importance of what they were doing."

Fortunately, NASA had the experience and the resources to make that happen.

NASA's Space Flight Awareness program dates back to the days of America's first manned space missions. It aims to help everyone who works on even the most menial or minuscule part of the program understand that their job is directly linked to the success of the space program and the lives of the men and women who will be flying into space. This kind of public outreach is second nature to NASA and is deeply engrained into the agency's culture.

At Associate Administrator Bill Readdy's direction, NASA deployed Space Flight Awareness representatives to Lufkin and to each of the collection centers to help coordinate community relations activities. Wednesday and Saturday evenings featured presentations from an astronaut—sometimes even a comedy show—at the camps. The astronauts were delighted to meet the searchers, and they gladly signed autographs late into the night.[1] Other NASA personnel carried photos, pins, decals, and patches everywhere and handed them out freely. It was a wonderful way to reach out to the people who were sacrificing so much to be part of the recovery.

The collection centers also began showing the fire crews some of the items recovered during the searches and explaining how each piece was significant.[2] "You're going to make the difference in us figuring out what happened and get us back flying again," Nacogdoches collection center manager Stephanie Stilson told the searchers at her camp.

The education went both ways. Mike Ciannilli met a father and son on one of the crews. After they talked for a few minutes, he thanked them for their help. The father said, "My son is making more of a sacrifice than me. He just got married. This was supposed to be his honeymoon." When the son heard about the accident, he told his new bride he felt compelled to help his country. She agreed it was important for him to do so, even at the expense of canceling their honeymoon.

"They weren't bragging about it," Ciannilli said. "That's just the kind of people you met."

Gerry Schumann befriended a young Native American man who sought him out to talk about NASA and the space program every evening. The young man broke his leg about one week into his deployment. The crew's rules required he be sent immediately back home, and he was heartbroken about having to leave early. Schumann looked for a way to assuage the man's grief—some token to thank him and encourage him. Schumann gave the young man his hat, which was adorned by pins collected in swaps with various search teams, and told him, "Here—you earned this." The young man was intensely moved by the gift. No one had ever done such a thing for him. When Schumann arrived home from his assignment in May, a package was waiting for him—a decorative Native shirt made by the young man's mother.

The communities also showed their appreciation to the NASA family by treating them like royalty. Jerry Ross served as Grand Marshal at the April 19 Annual Lufkin Downtown Hoedown parade, in which NASA's Mike Ciannilli and United Space Alliance chief engineer John Cipoletti walked with the cattle. Ross also opened a rodeo in Palestine and helped plant a grove of *Columbia* memorial trees in Nacogdoches.

The City of Nacogdoches asked the Space Flight Awareness representatives assigned to the city's staging area to serve as masters of ceremonies at the town's Special Olympics. "I can't tell you how many people came up to us at the end of the day and expressed their appreciation," Jim Furr said. "I was just a nobody—but to them I could have been the NASA administrator."

The City of Nacogdoches also requested that NASA provide them with a *Columbia* banner to lead in the national colors at the opening ceremonies of a rodeo. Furr said, "When all those beautiful horses came flying in, and all the gorgeous girls in their sparkly cowgirl outfits, and then they dimmed the lights and had a floodlight on the *Columbia* flag, I just lost it. It was a rough evening for me emotionally."

Mailbags filled with letters of appreciation, support, and encouragement arrived at the collection centers and at Johnson and Kennedy. They were unanimous in their sentimental messages. *We're sorry for your loss. We're praying for you. We believe in you. Keep flying.*

—

By day, the searches continued at a relentless pace. Helicopter overflights produced excellent results, but flying low enough to spot debris from the air exposed the helicopter crews to extreme danger. Low-and-slow flying put the helicopters well within the "dead man's curve"—a combination of speed and altitude in which a crash was unavoidable if a chopper's engine suddenly quit. With enough altitude and airspeed, a quick-thinking helicopter pilot can rapidly descend, using the resulting relative wind to maintain control and "autorotate" to an unpowered landing. If the engine shuts down when the helicopter is too low, there is insufficient time for the pilot to react. Little can be done to keep the rotors turning to maintain control.[3]

The low light of early mornings and late afternoons, and the hazardous environment of much of the search area, occasionally produced minor incidents and near accidents.[4] Mike Ciannilli said, "There were power lines out in the middle of nowhere for no reason—wires hanging between trees, wires between abandoned farm buildings. You never had a clue they were there. You'd find them at the last second, and then suddenly you're in an emergency situation."

In late March, one helicopter pilot misjudged the terrain in which he was landing and damaged his helicopter. Ed Mango said, "No one was hurt, but that started us thinking that maybe we shouldn't be landing to pick up things."

Luck ran out just a few days later, on the afternoon of March 27.

Pilot Jules "Buzz" Mier was flying his Bell 407 helicopter over the Angelina National Forest, near the town of Broaddus and the Sam Rayburn Reservoir. Mier was a Vietnam veteran with more than one thousand hours of combat flight experience and more than eight thousand hours of overall flight time. He had served as a flight instructor at Fort Rucker, Alabama, while also flying Medevac missions with the Alabama Air National Guard. Most recently, he had been flying tourists around the Grand Canyon.

Sitting up front with Mier was search manager Charles Krenek of the Texas Forest Service. Krenek had twenty-six years of experience as an aviation specialist and a wildland firefighter. A resident of Lufkin, Krenek was well-known and well liked by the forest service community in East Texas.

In the back of the chopper were three members of the search party. Matt Tschacher was with the US Forest Service from South Dakota. The technical experts from KSC were Richard Lange, a space shuttle fuel cell cryogenics support worker with United Space Alliance, and Ronnie Dale, with NASA's Safety and Process Assurance Branch.

After stopping for lunch and refueling, the crew took off at 3:15 p.m. on their second mission of the day. About an hour into the flight, just barely above the treetops, the helicopter developed a problem. William Dickerson, a local resident, was on a fishing trip in the vicinity with his nephew. They saw the helicopter fly over, and its engine suddenly went silent.[5]

The helicopter hit nose first into the crown of a large oak tree in swampy Ayish Bayou.[6] The cockpit was crushed, killing Mier and Krenek instantly. All of the men in the rear of the helicopter were seriously injured—but they survived.

Dickerson and his nephew found the helicopter's wreckage. They helped the three survivors out of the swamp and to the side of a nearby road, and then went to call for help.

Doug Hamilton from the US Forest Service and Sheriff Tom Maddox were among the first people on the scene after Dickerson's phone call. They met the three injured men and learned what had happened. Hamilton and several other men waded back into the woods, through water that was several feet deep in places, to find the partially submerged helicopter. They located the bodies of Mier and Krenek in the wreckage. There was no way to bring them out of the woods and the swamp without assistance.

Marsha Cooper and Felix Holmes heard Hamilton's call. They brought a bulldozer and several all-terrain vehicles as close as they could to the accident site. By knocking over pine trees with his bulldozer, Holmes built a makeshift path for the four-wheelers to reach the scene.

The accident dealt a devastating blow to the Texas Forest Service, the US Forest Service, and the East Texas community. Residents could scarcely fathom that one of their favorite sons had given his life in the search for *Columbia.*

Crowds packed Krenek's funeral service in Lufkin. One attendee estimated nearly one thousand people were on hand, in a church built to hold perhaps four hundred.[7] NASA's astronaut corps was well represented. They understood the magnitude of the sacrifice made by these men and their community for the space program.

Having attended the memorial services for their colleagues on *Columbia*'s crew just one month earlier, the astronauts could not help but observe how the families of the fallen crewmen and the community reacted to the accident. Dom Gorie noticed Krenek's wife Charlotte singing during the service, and he met her afterward. He was moved by the strength of her faith in such a tragic circumstance. Gorie said, "If she could stand up and sing at a service like that after losing her husband, it gave you confidence to do anything. If somebody can endure that and press on, we could certainly press on with whatever task that was put in front of us. It was powerful."

Gorie also noted that no one appeared to be voicing regrets or blame. No one believed that these men lost their lives doing something insignificant. Rather, the community regarded it as a sacrifice to an important undertaking. It proved that the people of East Texas were doing their utmost to help NASA return to flight, no matter the personal cost.

Now that they had given the lives of one of their own men to the cause, it was more important than ever to ensure that the task was worthwhile—that these two men and the crew of *Columbia* had not died in vain.

The motto that had been circulating since the early days of the recovery effort now seemed even more poignant. *Their mission became our mission* took on a much deeper meaning for the citizens of East

Texas. They were now inextricably part of the *Columbia* story—their own blood mixed with the blood of NASA's astronauts.

Ed Mango, astronauts Jerry Ross, Dom Gorie, and John Herrington, Dave Whittle, and a representative from FEMA attended Buzz Mier's funeral service several days later in a chapel on the rim of the Grand Canyon. Ross spoke at the service and gave items of appreciation in tribute to Mier's family members. It was a fitting commemoration for a man who had dedicated his life to serving his country.

The crash caused an immediate stand-down in air operations for an investigation of the accident. The NTSB recovered the helicopter and tested its engine. The problem was traced to a failure of a component in the fuel control unit. Fuel had just stopped flowing into the engine. Contributing to the accident was lack of a suitable place for Mier to make a forced landing.[8]

Many of NASA's searchers were reluctant to fly in helicopters after the accident. They felt it was too dangerous. The NTSB reminded the leaders in Lufkin that it made no sense to risk people's lives in a debris salvage operation. FEMA suggested terminating the air search efforts altogether.

NASA was reluctant to abandon the air operations, though. The leadership team eventually worked out a compromise that would enable them to continue helicopter searches safely. First, flights were to be at a higher altitude, to give the pilots time to recover from emergency situations and find a clear place to land. Second, there would be no more landings to pick up debris. Spotters were to note an object's position and call in a ground search team. Finally, operations would be concentrated farther to the west in the debris field, where there was less forest, and where debris from *Columbia*'s left wing and aft structure was more likely to be found. With these changes made to the mission profile, air searches resumed on April 10.[9]

The accident also resulted in three safety recommendations from the NTSB to the FAA addressing the problem that caused the crash. As a result, no further crashes of Bell 407 helicopters have occurred worldwide due to the same type of failure.[10]

Even with the temporary stand-down in helicopter flights, the search effort continued at an intense pace. Everyone was still incredibly busy, but gone was the sense of chaos from the first weeks of the recovery effort.[11] As time went by, the areas of interest in the ground search changed. The first six weeks' searches were focused in the area between Nacogdoches and Hemphill. By April, the emphasis had shifted to the area around Corsicana, where pieces of the shuttle's tile and wing leading edge materials were being discovered.[12]

The results were incredible. By April 7, more than sixty thousand pieces of shuttle debris—totaling 70,700 pounds—had been collected from the field and transported to Kennedy Space Center. Of the 137 tanks and cylinders with hazardous materials on board *Columbia*, 76 had been recovered. Ground searchers had covered 77 percent of the assigned acreage in Texas and Louisiana. Water searches in Toledo Bend were 99.5 percent complete. Aerial searches were on temporary hold following the March 27 accident, but 76 percent of the assigned aerial grids had already been searched.

The personnel staffing snapshot on April 7 showed 5,545 people working in search operations. This included 683 EPA, 75 FEMA, 282 NASA, and 4,289 US Forest Service personnel engaged in the ground search; air operations had fifty-five people assigned; and 130 people were working on diving operations. The Texas Forest Service had twenty-two people assigned, and there were a few other resources from DOD, DOT, and the Texas Department of Emergency Management.[13]

—

STS-114, the space shuttle's next planned mission, was on hold but not canceled. NASA did not yet know when the mission might fly—it could be a matter of months or years. In the meantime, astronaut Eileen Collins, the mission's commander, needed to keep her crew focused, engaged, and working as a team.

Although they could not commit to a long-term assignment, she hoped that her crew might be able to do something to help the

*Columbia* recovery effort. Collins asked Jim Wetherbee and Dom Gorie if the STS-114 crew could walk a search one day. They agreed that it would be a positive way to show the astronaut office's support and gratitude for the ground search. It would also demonstrate NASA's intention to fly the shuttle again as soon as possible.

Collins and her crewmates Jim Kelly and Soichi Noguchi flew into Nacogdoches Airport on Thursday, April 10. After a tour of the collection facility and a safety orientation, they went to a search site.[14]

They were told in advance about the physical conditions to expect in the field, but Collins was uncertain about what the emotional state of the search teams would be. She had attended all of the funerals and memorial services for the *Columbia* crew. She expected people working the recovery effort to be grim or sad. Much to her surprise, she found the search crews and her NASA colleagues upbeat. "I realized that you just can't be looking at a funeral all those weeks," she said. "You've got to start living again."

Collins, Kelly, and Noguchi walked with searchers for about two hours. Noguchi found a piece of tile. Collins felt somewhat disappointed not to find anything herself, despite being intensely focused on the ground around her. She thought it might have helped her gain some measure of closure with her friends who perished on *Columbia*.

Collins and crew then flew to Lufkin. That evening, they gave a presentation at the Civic Center for the Lufkin community, showing videos about the space shuttle and the International Space Station. The crew distributed stickers of their mission patch and signed photos for the attendees. Once again, it was a way for NASA to thank the residents of East Texas for their extraordinary support and sacrifices in the cause of manned spaceflight.[15]

—

On April 23, five *Columbia* crew spouses came to East Texas to thank the recovery teams and to visit the area where the crew's remains were found. Jan Amen of the Texas Forest Service was one of the drivers for the group. After they toured the Nacogdoches collection center, Amen escorted Lani McCool and Rona Ramon to the Hemphill area. FBI special agents Terry Lane and Ed Zalomski accompanied them. The other spouses stayed in the Nacogdoches area to meet with some of the recovery teams.[16]

The landowner on whose property Ilan Ramon's remains were found had been informed that Mrs. Ramon was coming to see the site. In preparation for her visit, he built an access road through the woods to the recovery area—at his own expense—to ease her journey.[17]

Brother Fred Raney met with Mrs. McCool and Mrs. Ramon and told them how their loved ones had been cared for in the "chapel in the woods." Brother Fred said, "I wanted them to know that they were being thought of during that whole time."

Jan Amen wrote to a friend of her experience that day: "Who gets to do that? I was so humbled by that honor."

—

The navy wrapped up its operations on April 13. The only two *Columbia* debris objects retrieved from Toledo Bend Reservoir were the landing gear brake assembly found by the FBI on February 10 and the thumbnail-sized piece of shuttle skin found in March. Dive teams were unable to locate any debris in Lake Nacogdoches.

The navy knew going into the operation that the cards were stacked against their finding anything in the reservoirs. Despite the challenges, the navy brought its highest skilled personnel and most sophisticated search equipment to the scene and worked continuously for two months.

The navy concluded that despite the "ear witness" reports, it appeared that no large, intact sections of the shuttle landed in the lakes in the primary search area. Smaller pieces may have fallen into the lakes, but they could not be located with current technology and within time and resource constraints. And short of draining the lakes—which was clearly impracticable—it was impossible to find smaller debris.[18]

In a sense, it was a good thing that the navy did not find shuttle material in the lakes. NASA could be relatively certain that nothing toxic or of potentially significant value to the investigation was still underwater.

The remarkably efficient ground and air searches cleared the debris field on schedule. By mid-April, Dave Whittle said, "It was becoming obvious that we were not finding the big parts anymore, and the smaller stuff probably provides less information than the big stuff does. And we were also looking at our maps and seeing we were about finished."

It was time to begin closing down the search effort.

One of FEMA's three primary objectives for the operation was to return the disaster area to the condition it was in prior to the accident. Whittle and his colleagues met with each of the Texas county judges from the affected counties to ensure they felt that things had returned to normal. Whittle also made certain the judges understood how to reach NASA's *Columbia* Recovery Office at KSC if anything else turned up in their communities.

On April 18, Palestine became the first camp to close. A few days later in Hemphill, Choctaw firefighters from Oklahoma staged a Native fellowship or victory dance in the VFW hall to celebrate the successful completion of the search in Sabine and San Augustine Counties. Then Hemphill's camp and collection center closed on April 22.

The place felt eerily silent after three months of nonstop activity. After ensuring that everything had been cleaned up properly, Greg Cohrs went home and became sick for nearly a week. Just as had happened up to the point that the last of *Columbia*'s crew members were recovered, Cohrs said, "My exhausted and drained body was able to resist illness until I completed the work."[19]

On Pat Adkins's last day in Hemphill before he returned to KSC, he walked around the downtown area and visited every shop. He tried to buy something in each shop as a small way to say "thank you" for the way everyone in town had taken such good care of him.

Search operations continued for a short time in the western end of the debris field, until the number of identifiable shuttle pieces being recovered dropped to less than one per grid.[20] At this point, the CAIB had already announced their conclusions as to the root cause of the accident. Additional material recovered was not likely to change those conclusions.

All ground search operations in Texas ended by April 30, and the remaining fire crews demobilized on May 1. The Nacogdoches site closed down on May 3, and Corsicana closed on May 4. The Longview staging area shuttered its operations on May 7.[21]

Limited search operations moved to Utah starting on May 2. Teams spent about eight days trying to find an object tracked by radar after falling off *Columbia*. They were unable to locate the object or any other debris from the shuttle in their search.

—

NASA's Space Flight Awareness organization sponsored a huge dinner in Lufkin on April 29 to celebrate the end of recovery operations.

The event was held in Lufkin's Civic Center, where the emergency operations center was established on the afternoon of the accident. Huge posters from KSC and the recovery sites hung on the walls. Tables and chairs were draped with gold covers. The scale of the event was impressive—a party only Texans can throw—feeling to Ed Mango like the celebration scene in the movie *The Right Stuff*.

Jan Amen reported, "Dinner was steak and chicken, green beans, rice, rolls, salad, pie, all prepared by the Diboll Country Club. Free drinks flowed freely."[22]

Administrator O'Keefe hosted the event for NASA, and Scott Wells spoke on behalf of FEMA. County judges and civic leaders from every county in East Texas were on hand. Judge Jack Leath, Tom Maddox, Greg Cohrs, Roger and Belinda Gay, Marsha Cooper, and a host of other people represented Sabine County. Dignitaries from the various Native American Tribes and Nations attended. An astronaut sat at every table.

Dom Gorie opened the ceremony with a heartfelt invocation that brought tears to the eyes of nearly everyone present. The Expedition Six crew sent a live video message from the International Space Station. A video about the *Columbia* crew followed.

O'Keefe and Dave King presented plaques recognizing the nation's appreciation for the contributions of the people at every table in the hall. The spouses of *Columbia*'s crew spoke of their gratitude to the people of East Texas for bringing their loved ones home again. Eileen Collins closed the ceremony on behalf of the next shuttle crew scheduled to fly in space.[23]

It was a fitting and emotional close to a tumultuous three months. The people of East Texas had provided the nation and the world with an enduring lesson in how to handle a crisis with dignity, compassion, and competence.

At the end of the evening, after Jan Amen dropped off her last load of astronauts and families at their hotels, she emailed to a friend, "I absolutely lost it. I squalled all the way back to Cudlipp like a big fat crybaby. I'm whooped!"

—

The largest land search-and-recovery operation in United States history had finally ended. This was the first incident under the overall auspices of the new Department of Homeland Security, and about 450 federal, state, and local agencies and volunteer organizations worked together in a textbook example of interagency cooperation and collaboration with local communities.

In the weeks between February 1 and May 10, 2003, nearly twenty-five thousand men and women searched 680,750 acres of land—in essence, walking every square foot of an area roughly the size of the state of Rhode Island. The US Forest Service, Bureau of Indian

Affairs, Bureau of Land Management, National Park Service, US Fish and Wildlife Service, and state forestry organizations and contractors provided most of the searchers, particularly after February 17.

Aircrews logged five thousand flight hours and painstakingly searched 1.6 million acres, an area nearly 50 percent larger than the state of Delaware. The helicopter searches were crucial in determining what caused the *Columbia* accident. Helicopter spotters located 65 percent of the 2,700 shuttle components that eventually ended up on the grid of the reconstruction hangar.[24]

Divers from the navy, FBI, Houston police department, EPA, Texas DPS, and the Galveston police department conducted more than three thousand dives and spent more than eight hundred hours on the bottom of lakes searching for debris from *Columbia*.[25] The overall water search effort covered twenty-three square miles of lake bed.

The combined efforts of these remarkable women and men totaled 1.5 million man-hours.[26] They recovered nearly eighty-four thousand pieces of *Columbia*, with a combined weight of 84,700 pounds. That was equal to about 38 percent of the shuttle's landing weight. Some of the most critical pieces recovered included the OEX recorder, more than 90 percent of the crew module, and pieces of the heatshield and structure from the left wing and the left side of the orbiter. Every piece of debris recovered was cataloged. This material would provide vital information on how the accident occurred and how the shuttle's structure was affected as the vehicle broke up.[27]

Most emotionally important to the NASA family, the remains of all of *Columbia*'s crew were recovered from the field and returned to their loved ones for interment.

The question remained, though: *Could the space shuttle fleet be made safe enough to fly again?*

## Chapter 11

# RECONSTRUCTING *COLUMBIA*

Only a few days into the recovery period in early February, we realized we needed a place to lay out *Columbia*'s debris in order to reconstruct the accident. NASA's management selected Kennedy Space Center as the best place for that to happen. The people chosen to put *Columbia* back together—an unprecedented task—would be the technicians and engineers at KSC. They were the people who had cared for her daily for the previous twenty-three years. The reconstruction ended up being a collaborative effort of many NASA centers, but Kennedy was the lead.

Even though I was selected to head the overall reconstruction effort on February 9, things were already well under way at Kennedy when I returned home from Barksdale a few days later. This was thanks to the hard work of shuttle test director Steve Altemus. Altemus happened to be walking by an office on February 3 when he overheard someone being offered—and then turning down—the role of *Columbia* reconstruction director. Altemus volunteered on the spot. He had fifteen years of experience in all aspects of shuttle launch, operations, and landing, as well as emergency management. He was the perfect person for the job.

We divided our duties, with me as the "up and out" manager—dealing with the reconstruction effort's interfaces to management, the CAIB, the press, congressional visitors and other VIPs, and other NASA centers—and Altemus leading the "down and in" day-to-day operations. We knew each other well, having worked together for many years. We made a strong leadership team with complementary skills.

—

The Reusable Launch Vehicle hangar at the southeast end of the Shuttle Landing Facility was only being used for storage in early 2003. It was the perfect size, secure, and convenient. Although it was on NASA property, the state of Florida owned the hangar. We quickly secured a short-term lease as soon as we got the "Go" to stage the reconstruction at Kennedy.

Getting everything set up in less than two weeks, before the first truckloads of debris were due to arrive, was not going to be an easy task. The hangar's leaky roof needed to be fixed. Processes for receiving, decontaminating, categorizing, and cataloging the debris had to be developed. Someone had to build an area for reconstructing the crew module. Much to Altemus's relief, KSC director Roy Bridges promised him the unconditional, immediate support of all of KSC's organizations for whatever he needed to ensure the reconstruction happened smoothly. Altemus would not have to contend with any bureaucratic hassles.

The technical term for the process the NTSB proposed was "reconstruction," but this was not going to be an attempt to rebuild the ship from her wreckage. Rather, we would lay out the recovered debris on a floor plan representing the shuttle peeled open onto a two-dimensional grid, with the shuttle's entire outer surface and the wing structures facing upward. By placing the recovered pieces of the shuttle next to one another, as they would have been on the ship, investigators could look for patterns in the debris that might indicate how *Columbia* was damaged and came apart.

Under NTSB's guidance, NASA and Boeing Air Safety used yellow tape to lay out a dimensioned grid on the hangar floor, with a blue tape outline of the shuttle's sections superimposed over it. The

NTSB suggested that the outline and grid be 10 percent larger than the actual shuttle. Additionally, the major component areas were separated from each other by about three feet. This would provide room for people to walk around between structures and examine the debris from all angles. The blue tape outline was oriented as if the shuttle had been pulled into the hangar nose first. From there, the outline was drawn as if the shuttle was turned belly side up and splayed out into two dimensions. Each wing was outlined in three separate sections: lower surface tiles, lower surface structure, and upper surface tiles. The idea was that when enough debris was available, placing the lower surface tiles over the lower surface structure would enable engineers to study those components in contact with one another.[1]

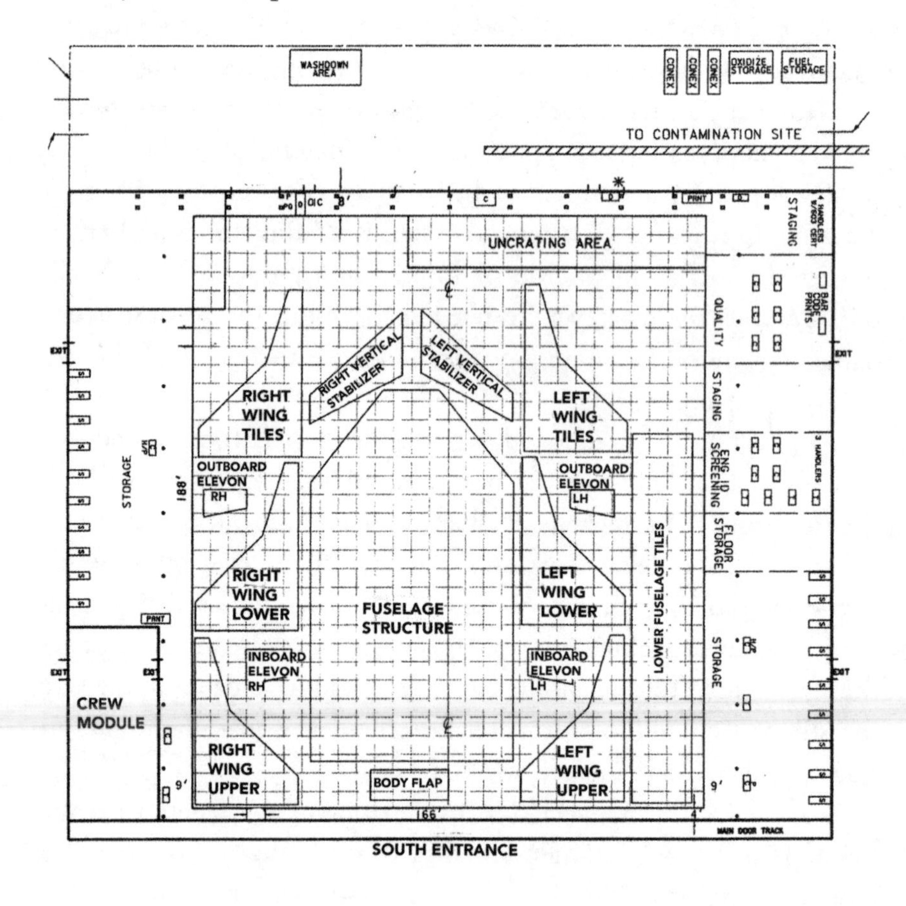

Laying out the shuttle on the grid made it easy to see patterns—burns, cracks, stress marks, and the like—in the pieces of the orbiter structure. Those could hint at the state of the vehicle when it broke up. Items from the external surfaces were to be placed on the grid. All materials from the crew module—the compartment in which the astronauts were sitting for launch and landing—would be reconstructed in three dimensions inside a walled-off area with controlled access. Debris from the shuttle's internal systems—wiring, plumbing, avionics, and so forth, as well as materials from the modules in the payload bay—would go onto "bread racks" along the walls of the hangar. These items were thought not to have contributed to the accident, so they were segregated from the structural debris but readily accessible if necessary. Also lining the hangar walls were workstations for the various engineering disciplines to examine pieces of debris as they arrived from Barksdale.

We set up a separate decontamination site outside of the hangar to protect its workers from potentially hazardous materials. It was possible that some debris might arrive from Texas still contaminated with hypergolic fuels. Pyrotechnic charges might also be attached to some of the wreckage, and these would need to be handled off-site. There was a remote possibility of exposure to biological hazards, either from experiments on the Spacehab module or even from human remains that might have inadvertently made their way back to Florida in the debris. Fortunately, this did not turn out to be an issue, but we needed to have a process established for any contingency.

The NTSB suggested the personnel receiving the debris make a rough hand sketch of the appearance of every item as it arrived. Technicians could also write notes to be filed with the sketches. "That process quickly went out the window, as soon as the trucks started arriving," Altemus said. Neither NASA nor the NTSB had anticipated we'd recover eighty-four thousand pieces of debris—and most of them would be the size of a nickel. Sketching every item was simply too much work. However, we did photograph and bar code every item received—a level of detail far beyond that in typical NTSB investigations.

With the *Columbia* Accident Investigation Board's concurrence, we eventually opened up a second location in an equipment hangar adjacent to the Apollo Saturn V Visitors Center across the road from the Shuttle Landing Facility. This building held friable materials that posed a health hazard because of their microscopic fibers. Some of the items placed in this building were Kevlar-wrapped pressurized tanks and pieces of the payload bay doors.

—

KSC engineers and technicians were eager to work in the hangar. The people who had tended *Columbia* over the years desperately wanted to help out with the accident investigation in some way. Many of them had volunteered to be deployed to East Texas to help in the search, but for one reason or another, their management had told them that it might not be the best use of their talents. Despondent over the loss of the shuttle, they were frustrated about the lack of opportunities to do something "useful" in the aftermath to take their minds off their grief. Working in the reconstruction hangar was not going to be an easy task. However, it would be a welcome chance for people with specialized knowledge of the shuttle's systems to use their talents to solve the mystery of what doomed *Columbia* and her crew.

Orbiter Project Office head Ralph Roe asked Jon Cowart to be the NASA engineering lead from Houston for the reconstruction. Warren "Woody" Woodworth was designated as the chief *Columbia* reconstruction engineer for United Space Alliance (USA). Jim Comer helped Steve Altemus set up the hangar and then became the head of operations at the hangar for USA, overseeing about three hundred contractors in the reconstruction.

The Astronaut Office assigned Pam Melroy to manage the crew module reconstruction, because she was the lead "Cape Crusader"—the astronauts assigned to work on shuttle issues at KSC. Melroy and her colleague Marsha Ivins alternated workweeks in the hangar, until Ivins moved on to other duties about a month after the accident. Melroy and her staff of six people would have the unenviable task of examining what was left of the crew module and the equipment used by the astronauts. Two of her key assistants were John Biegert and Robert Hanley.

Biegert, a long-term member of the flight crew systems engineering group from Houston, supervised the teams that loaded all of the crew equipment, lockers, food, EVA suits, experiments, and other items—five thousand to seven thousand pieces of equipment and materials altogether—into the crew module in the ten days prior to *Columbia*'s launch. He and his teams knew where every item should have been stowed when *Columbia*'s crew prepared the ship for landing.

Robert Hanley—one of the last people to see *Columbia* in flight—had been going "stir-crazy" in Houston and was desperately looking for something useful to do. He asked permission to visit KSC and see the reconstruction effort. He only intended to be there part of the day and return on the afternoon flight back to Houston. He walked into the crew module area and saw Pam Melroy writing on a whiteboard. Assuming that he had come to help identify debris, she said, "Robert! Great! How long can you stay?" He replied that he just needed to go back to Houston and get some clothes, and then he could stay as long as she needed his help. He became Melroy's lead engineer.

Renée Ross was an example of someone brought in for specialized expertise. Ross, a USA employee who managed *Columbia*'s flight-data books, received an email from Melroy, asking her help in identifying some fragments of pages that appeared to be from various manuals in the crew module. After a few weeks of exchanging photos of documents via email, Melroy asked Ross to come to KSC to work through the accumulated backlog of crew module document remnants that were being returned from the field. Ross eventually identified and cataloged about four hundred pieces of documentation from *Columbia*.

Thermal protection system lead engineer Ann Micklos, who had been dating *Columbia* crewman Dave Brown, received a call from her chief engineer. Micklos recalled, "Knowing my circumstances, he told me, 'We really need you to be part of the reconstruction crew. You know *Columbia*'s systems and you're the right person to be there. Is this something you think you can do?' Well, it's human nature to want to help during a time like this. There was no way I could just sit in my office. Obviously, there was no place I'd rather be than with *Columbia*, even with how difficult the challenge initially was."

Finally, some of the "founding fathers"—the original designers of the shuttle at Rockwell International, now a heritage company within Boeing—came to the hangar to help. John Tribe and Sam Kreidel were among the senior Rockwell engineers who had lived with the shuttle since 1972. They knew how the shuttle was originally designed and built.

In contrast, the engineers and technicians from Kennedy knew the current status of the systems and components some twenty-odd years later. There were no digital drawings of most of the shuttle's components, so the combined knowledge of the designers and the hands-on staff was needed to identify some pieces of debris.[2] For example, the outboard tires on the left main landing gear could only be specifically identified by the presence of balancing patches that were installed at KSC.

All told, roughly four hundred scientists, engineers, and technicians worked in the hangar during *Columbia*'s reconstruction. Most of the workers were from United Space Alliance (who maintained the shuttles at KSC) and Boeing. Nearly every NASA center was represented.[3] Operations ran in two shifts per day, six days per week.

—

Several of the people who reconstructed *Columbia* had also worked with *Challenger*'s wreckage following the 1986 accident. Even though the circumstances of the accidents and the reconstruction processes were vastly different, some lessons learned from seventeen years earlier assisted in the *Columbia* reconstruction.

*Challenger* fell into the ocean without having left the atmosphere on her mission, traveling a little less than twice the speed of sound at peak velocity during ascent. Despite the way the accident appeared on television, the shuttle stack did not explode. Rather, the vehicle broke up because of aerodynamic forces after the external tank structure failed. Most of *Challenger*'s debris—much of it in large chunks and sections—came down in a relatively confined area within sight of the Florida coast.

*Columbia*, on the other hand, had been traveling in excess of Mach 18 at an altitude of over two hundred thousand feet when it disintegrated. Its wreckage was twisted, shredded, subjected to plasma, melted, oxidized, burned, and scattered over a 250-mile-long path. The vast majority of the debris that came back from *Columbia* was smaller than an office desk. Much of it was the size of a nickel.[4]

NASA's primary goals in the *Challenger* recovery had been to retrieve the crew, the right-hand solid rocket booster (SRB) with the O-ring that had burned through and caused the accident, and the crew module. NASA also was interested in debris from the shuttle's payloads, the left-hand SRB, the external fuel tank, and a few other specific components. The accident investigation was not focused on the orbiter itself, because the vehicle was clearly not the cause of the accident. There was no need to document the latitude and longitude of where each item was found.

Navy ships and divers spent seven months scouring the sea floor for the priority pieces of *Challenger*'s debris. They retrieved 50 percent of the SRBs and less than 50 percent of the orbiter and external tank.[5] About 70 percent of the crew module and the surrounding fuselage structure was recovered.[6] The remains of *Challenger*'s astronauts were inside the crew module when it was recovered. The crew compartment had been in the water for more than a month before the navy located it.

Some parts of *Challenger* unrelated to the accident were left on the seafloor rather than incurring the risk and expense of retrieving them.[7] Pieces of *Challenger*'s wreckage occasionally wash up on the Florida beach to this day, decades after the accident.

NASA reconstructed parts of *Challenger* in KSC's Logistics Facility, about one mile south of the VAB on Contractor Road. I clearly remember it. I was working in design engineering and launchpad safety and structural systems at the time, and I was one of the employees permitted to see the debris during *Challenger*'s reconstruction. My first impression was the horrible smell. *Challenger* reeked of seawater and rotting barnacles. A few of the shuttle stack's potential failure points were laid out on the floor, including the SRB aft field joint that burned through and the attach point from the SRB to the external tank. A few sizable pieces of the fuselage were propped up using two-by-fours in a rudimentary step toward reconstructing the ship. Much of the smallest debris was basically swept into piles off to one side. The crew module debris reconstruction took place in a small building adjacent to the Logistics Facility, with tightly controlled access.

John Biegert worked as part of the reconstruction teams for both *Challenger* and *Columbia*. He said that perhaps the biggest difference he saw between the two efforts was due to the advancement in digital technology over the course of seventeen years. Instead of using Polaroid cameras and the very limited computer technology of 1986 to document the debris, the *Columbia* accident investigation benefitted from digital photography, three-dimensional scanning and modeling, sophisticated computer databases, and other relatively new technologies.

—

The *Columbia* Accident Investigation Board wanted to maintain an on-site presence at the reconstruction hangar. I worked with them to determine where their offices would be located and how they would interact with the debris and the workers at the hangar. The CAIB set up in office trailers parked outside the hangar. This enabled them to observe what was going on, provided them private workspaces, and kept the Board from distracting our engineers and analysts who were trying to do their own work. This was not done to protect the information—it was to protect our people from having too many cooks in the kitchen.

We had an interesting relationship. If the CAIB wanted something, we were obligated to give it to them—but they could not direct our team. They could make requests, and we would fulfill the requests. The process seemed slow to them, but we did not drag our feet. We were doing it at the pace we were used to at KSC, which was meant to be methodical.

Greg Kovacs, a medical doctor and professor of electrical engineering at Stanford, came in toward the end of February as an advisor to the CAIB. He spent almost all of his time in the hangar with us, and he became a good friend and de facto member of our team. He jokingly observed that the "NASA approach to things is to put 50,000 people in a line and move forward an inch at a time."

It did seem painstaking and slow sometimes, but we were following a basic tenet of the NTSB's investigation procedures—to avoid speculating about the cause of an accident for as long as possible. The NTSB's long experience proved that someone whose mind is latched onto a given theory will pursue that line of investigation and disregard evidence that points to other possibilities. That could lead to dead ends and wasted time. In the long run, the NTSB said, keeping an open mind as long as possible would actually speed up the investigation.

We learned and recited the NTSB mantra: "Keep an open mind. Let the debris tell the story. Collect the debris, and the debris will tell you what happened."

In the end, the investigation concentrated on the problem areas identified in the fault tree. We knew the problem did not start on the right wing, so we didn't spend a lot of time in detailed examination of the right wing. We knew where we had to concentrate our efforts, but being able to compare *Columbia*'s right wing to its left wing was invaluable throughout the process.

—

The NTSB was initially concerned that our workers might be overwhelmed in trying to deal with the quantity of wreckage arriving from the field. However, this turned out not to be an issue. Because Steve Altemus set up and staffed the hangar and its receiving processes so quickly after the accident, the facility and its personnel were ready to go as soon as the first two truckloads of debris left Barksdale on February 12 and arrived at KSC on February 13.

Barksdale dispatched a shipment to us whenever enough material filled a flatbed or an enclosed eighteen-wheeler. In the first several weeks, shipments arrived at KSC every other day. Just one month after the arrival of the first truckloads of material from Barksdale, the hangar already contained 33,798 pieces of debris, totaling 43,200 pounds, and representing more than 19 percent of the shuttle's dry weight. The frequency of shipments began to tail off slowly as recovery operations cleaned out the debris field—first to a couple of shipments per week, and finally only one truckload per week.

KSC security cars escorted the shipments, and NASA special agents alerted state police along the route that the trucks were coming through. All of the NASA centers asked to participate in the honor of escorting *Columbia* back to KSC. To share that solemn role among the centers, one special agent from KSC and one from one of the other centers usually staffed the escort vehicles. These special agents also personally carried any sensitive materials being sent to KSC, rather than leaving them in the back of a truck. It was an eighteen-hour drive from Barksdale, with stops only for food and fuel.

KSC security special agent Linda Rhode (whom the reconstruction team nicknamed "Agent 99") accompanied one shipment back from Barksdale with her boss, Mark Borsi. "It was four o'clock in the morning, and it was pitch-black. I was driving, and we'd been spending probably fifteen hours staring at the back end of this truck," Rhode said. "Somewhere along Interstate 10 in the Florida Panhandle, a law enforcement vehicle jumps between me and the truck and pulls it over." Borsi got out of the car and told Rhode to stay put. She said, "The agricultural inspector was talking to the truckers and then got a radio call. I think he realized the error of his ways. He hopped in his cruiser and took off."

"Truck day" was always a special day in the hangar. It was not quite a celebration, but every load of new material meant that we were getting closer to figuring out what happened to *Columbia*. The staff and I went out to greet the drivers and thank them for bringing the material to KSC. Drivers told us where the material had come from. For some personnel in the hangar, this was their primary news source about events in the field.

As the truck was unloaded, every item was "sniffed" to ensure that it was not contaminated by hypergolic propellants. Staff was on the lookout for crushed tiles and other friable items whose fibrous materials could cause respiratory problems.[8] The processing centers in East Texas did their jobs well, because none of the material was contaminated by the time it arrived at KSC.

Materials were then triaged just inside the hangar from the parking lot. Every item arrived bagged (if it was small enough) or wrapped in cling-wrap plastic film, and carried a tag with an identifying number and the GPS coordinates where the piece was found. Quality assurance staff bar-coded, photographed, and cataloged each item into the database. Items that were related to payloads, fuselage, or internal structural items went to the appropriate engineering stations. Things that were obviously crew-related and debris from the crew module went directly to the crew module room. Other debris was examined to try to determine where on the shuttle it might have come from. Then it went into the bread racks or onto the floor grid as appropriate.

Identifying debris that may or may not have been from the shuttle proved an interesting challenge. One landing-gear strut was heavily oxidized and caked with mud when it arrived. "Before we got all that mud and crap off it, it looked like an anchor," said Jim Comer. After cleaning it up, many people believed it was from a B-52 bomber that crashed somewhere in Louisiana or East Texas many years ago. On

February 14, technicians pulled the endcap off to look for an inspection stamp inside the strut. Only then did we prove the piece to be from *Columbia.*[9]

Nearly half of the material that came back from the field was classified as "unknown" when it arrived at the hangar. These forty thousand pieces of *Columbia* were generally small, nondescript bits of hardware such as bolts, tubing, fittings, scraps of fabric, and wires. More than one dozen people in the hangar examined each unknown item to try to identify which system it might be part of. If the system could not be determined, then quality assurance classified the item by the material from which it was made—metal ceramic, tile, fabric, and so forth.

In a last attempt to identify the items, the reconstruction hangar held an "unknown party" in which all of the remaining unknown items were passed around for people to examine one final time.[10] By the end of the hangar's operation, only 720 items of the nearly eighty-four thousand pieces of debris remained formally classified as unknown.

—

The long association of KSC's staff with *Columbia* made them the most knowledgeable people about the ship. However, they were also the most susceptible to emotional reactions to the sad state of the vehicle. "This was like burying a friend," said astronaut John Herrington. "It wasn't just going to a wreck someplace and picking up pieces of an airplane that you don't have a connection to. These guys had touched every part of it in the processing facility or on the launchpad."

Workers reacted emotionally to the recognizable structures of the shuttle that came back from the field. Among the first pieces to arrive were the metal cockpit window frames. Those frames normally held three thick panes of aluminum silicate glass and fused silica. The glass was carefully polished and protected before and after each mission. The windows were tough enough to withstand collisions with small particles in orbit. But now, all that remained of the thick glass was a few small shards stuck in the frames, along with bits of grass and dried mud. "I was like, 'Oh my gosh! Why did this first piece have to be something that's so related to the crew? Why couldn't it be just a strut or something?'" said Ann Micklos. "You had seen the vehicle as a whole, and now you are viewing the parts in a manner you never dreamed of."

The mood on the hangar floor could be very somber at times. Every once in a while, someone would recognize a piece from a system they had previously worked on and then would break into tears. People looked out for one another and took their colleagues outside to regain their composure. "It wasn't just me, it was everybody," said Micklos. There was no way to predict when or where it would happen—but at some point, even the most hardened engineer or technician would break down when confronting a piece of wreckage and thinking about what it represented to them personally and to the crew.

One technician who had spent his entire career working on the shuttle's fuel cell power system was standing at the hangar entrance when a truck arrived with the broken, burned, and tortured cryogen tanks from his system. "He stood there in the hangar, looking out at the truck, and started to weep," said Steve Altemus. "He just said, 'Damn!' and walked away."

The crew module area was particularly tough to work in. Herrington described it as being "gut-wrenching," to stand among the remnants of seats, control panels, tools, and other items that his astronaut classmates had worked with on their mission. "Seeing the damage they'd suffered—and how little damage other items had incurred—really struck me. It really hit home," he said.

Our team leaders constantly monitored and tried to lighten the mood. When a large section of the air lock panel for the crew module arrived early in the reconstruction effort, people were concerned about how astronaut Pam Melroy would react to seeing this piece of the hardware. "I could see them looking at me, scared, thinking, *This is gonna be horrible*," Melroy said. "I just looked at them, and I could have kissed them! I said, 'Look at the size of this piece! You've brought me this great piece, and I'm so happy to have it back!' That set the tone, so that when they brought something in, it wasn't, 'Oh shit.' It was, 'Look at what we found for you!'"

No matter how hard people tried to keep the mood from becoming grim, things could change quickly and without warning. Jim Comer recalled examining with astronaut Marsha Ivins a contact lens case belonging to one of the crew. It was immaculate—without a scratch on it. Ivins laughed, "Hey Comer! This is what we need to build the next space shuttle out of!" And then a few minutes later, the two of them found the remnants of a cloth crew patch in which only the stitched border survived. The rest of the patch had burned away. Comer said, "We looked at each other and went, 'Are you *kidding* me?'"

We decreed that the hangar would be closed on Sunday—no exceptions. We knew that the staff would need a day each week to recover and recharge. "It's not a sprint, it's a marathon," Altemus said. "You've got to keep yourself healthy. That decision had a huge effect on the morale of the team, and we made fewer mistakes as a result."

On March 3, 2003, barely one month after the accident, former astronauts Wally Schirra and Jim Lovell came to KSC to encourage workers who were still grieving over the loss of *Columbia*.

Both men were well acquainted with the risks inherent in manned spaceflight. Schirra commanded Apollo 7, NASA's first manned mission after the fire that killed his friends in the three-man crew of Apollo 1. Lovell was commander of Apollo 13, when a deep-space explosion led to a harrowing several days in which the world watched anxiously and hoped that the crew would make it home alive.

Lovell said, "This is a risky business. Everyone I talk to says this should not stop the program—we should find out the cause."

Schirra encouraged KSC's team with Gus Grissom's famous line, "Do good work."

Lovell added, "We have a great program. Keep charging. Don't give up."[11]

The two astronauts also toured the reconstruction hangar. They thanked everybody for their devotion to the cause. It was as close to a pep talk as you can have in that kind of situation—almost like having your grandfather come and talk to you. It meant a lot to us.

Chris Chamberland and the KSC Web Studio produced a short video entitled *Sixteen Minutes from Home: A Tribute to the Crew of STS-107*. It was so powerful that as soon as I saw it, I wanted every hangar worker on all shifts to get a copy of the DVD. One morning in late February, I asked everyone to stop working and gather around a large TV in the hangar. We all sat on the floor, arms around each other, and watched as the crew did their thing in the video. At the end, I told everyone to pick up a copy of the DVD and then return to work whenever they were ready. We needed a break. The sharing of the grief and watching Rick and his crew enjoying themselves in orbit were really good. It was important to remember our friends in happy times.

With the passage of time, our staff eventually became somewhat inured to working with the broken pieces of the shuttle. Their depression gradually morphed into scientific and engineering curiosity about how the shuttle had come apart and what the debris was trying to tell them.

Even the pieces of wreckage that were not directly related to the accident held mysteries for us. Why did one piece of equipment come back heavily damaged, while another that was sitting right next to it was relatively unscathed? Why did all the propellant and other tanks in the ship come back in such good shape? Why were the oxygen feed lines in the engine manifolds more decayed than the hydrogen feed lines?

Solving these riddles engaged everyone's intellectual and engineering curiosity and kept us from dwelling too long on the tragedy represented by the debris of our beloved *Columbia*.

—

One day, a beat-up and torn stuffed dinosaur doll—yellow with purple polka dots—arrived in the crew module reconstruction room. Recovery workers in Texas thought there was a possibility that the crew might have flown the doll on the mission for a friend or family member. Robert Hanley knew what *Columbia*'s astronauts had taken on the mission. "Some things you absolutely knew could not possibly be crew-related stuff, and they'd go into what we called 'the East Texas trash box,'" he said. "We knew this dinosaur didn't fly, but we decided to keep it in the room as our little mascot—kind of a joke."

Pam Melroy cleaned up the room one day when Hanley was out of town, and she threw away the dinosaur. Hanley became upset when he returned and the dinosaur was gone. He scoured the trash boxes in the hangar and eventually located it.

Melroy felt guilty about inadvertently throwing away the doll, not realizing it meant so much to her team's morale. When the reconstruction effort was over, and she and her team returned to Houston, she had the doll repaired and dry-cleaned. Her crew module reconstruction team created the Yellow Dinosaur Club, made membership cards, and even crafted a badge for the dinosaur. Melroy flew it in space on the STS-120 mission she commanded in 2007.

"It sounds silly, but it goes back to what you were clinging onto to make this tragic thing something you could cope with," Hanley said. "The yellow dinosaur was just one of those things."

—

One of my roles was to carefully control press access to the hangar. Administrator O'Keefe gave me complete latitude on how to handle it. I wanted first and foremost to ensure that the press would approach the situation with the appropriate seriousness, and that they would not sensationalize the wreckage on the floor. Our workers' emotions were still raw after the accident, and the crew's families were still

Greg Cohrs (US Forest Service), Terry Lane (FBI), Greg Schumann (NASA), Debbie Awtonomow (NASA), and Olen Bean (Texas Forest Service) in the Hemphill VFW Hall on the one-year anniversary of the accident. *(Photo courtesy Gerry Schumann)*

Hemphill's debris collection center, in the county's Farmers Market Co-op shed. *(Jan Amen photo)*

Forester Rich Dotellis of the Texas Forest Service logs in debris recovered during the day at the Hemphill collection center. *(Jan Amen photo)*

Fire crews Florida 3 and 4 in San Augustine County on March 19, 2003—the day they found *Columbia*'s OEX recorder. *(Jeremy Willoughby photo)*

*Columbia*'s OEX recorder and tape reels—the "black box" that eluded searchers for forty-six days. *(Robert Pearlman/collectSPACE)*

Astronaut John Casper chats with members of a Native American fire crew at the Longview staging center. Casper, a veteran of four shuttle missions, was deputy director of NASA's Mishap Investigation Team. *(Jan Amen photo)*

A portion of an STS-107 Columbia Recovery Team banner, bearing thousands of signatures of searchers and other recovery workers who passed through Hemphill. NASA provided banners like this at every one of the staging areas during the recovery. *(Jonathan Ward photo)*

Charles Krenek's search crew poses for a photo while refueling on March 27, 2003, the day of the fatal accident. From left, Matt Tschacher (US Forest Service), "Buzz" Mier (pilot), Richard Lange (United Space Alliance), Ronnie Dale (NASA), and Charles Krenek (Texas Forest Service). *(Photo courtesy Boo Walker/Texas Forest Service)*

The STS-114 crew arrives at Nacogdoches on April 10, 2003 to spend the day with search crews. Left to right: Soichi Noguchi, Jan Amen (Texas Forest Service), Eileen Collins, Jim Kelly. *(Jan Amen photo)*

The reusable launch vehicle hangar (foreground), at the south end of Kennedy's Shuttle Landing Facility runway. It was here that engineers reconstructed *Columbia*'s debris. *(NASA photo)*

The interior of the reconstruction hangar on March 27, 2003, about two months into the debris recovery process. *Columbia*'s nose cap and nose landing gear are at bottom center. *(NASA photo)*

*Columbia*'s nose landing gear sits amidst other recovered pieces of the orbiter's structure in the reconstruction hangar, March 7, 2003. Engineering stations and "bread racks" storing miscellaneous components line the wall in the background. *(NASA photo)*

Reconstruction staff members gather around an STS-107 emblem on the runway apron outside the hangar. After this photo, the emblem was mounted above the hangar's sliding doors. I was later relocated to the Vehicle Assembly Building. *(NASA photo*

Jnited Space Alliance chief engineer Warren "Woody" Woodworth (red shirt) briefs *Columbia* :ccident Investigation Board chairman Admiral Hal Gehman (with glasses hanging around neck) and several congressmen in the reconstruction hangar. *(NASA photo)*

Portions of *Columbia*'s right-hand main landing gear and landing gear doors. Heavy oxidation of the landing gear strut, which was caused by chemical interaction with plasma during reentry, made it look like it had been rusting in the elements for years. *(NASA photo)*

Veteran astronauts Wally Schirra and Jim Lovell inspect the wreckage of one of *Columbia*'s elevon actuators on March 3, 2003. Left to right: Mike Leinbach, Lisa Malone (NASA Public Affairs), Jeff Wheeler (NASA Engineering), Steve Altemus, Jon Cowart, Schirra, Lovell. *(NASA photo)*

Cards and banners sent from students and well-wishers across America decorate the outer wall of the crew module reconstruction area. *(NASA photo)*

The yellow dinosaur that was the mascot of he crew module reconstruction team floats in *iscovery*'s cabin during Pam Melroy's STS-120 mission in 2007. *(NASA photo)*

Slumped and pitted tiles from the underside of *Columbia*'s left wing show that melted aluminum from inside the wing was spraying out onto the outer surface of the wing before the vehicle broke up. Despite their degraded state, these tiles protected the underlying metal of the ship's skin from melting. *(NASA photo)*

The tile table for the underside of *Columbia*'s left wing. Far fewer tiles were recovered from the left wing than the right wing, particularly aft of the leading edge of the wing in the center of the photo. *(NASA photo)*

Engineer Ann Micklos places a fragment on the left wing tile table in the reconstruction hangar. Each tile had a unique shape and was individually numbered. *(NASA photo)*

NASA structures engineer Lyle Davis attempts to reconstruct a part of one of the wing leading edge panels from small pieces of reinforced carbon-carbon found in the field. *(NASA photo)*

Technicians install a reinforced carbon-carbon leading edge panel on an orbiter's wing. This lustrates the size of a typical RCC panel and its underlying structure. *(NASA photo)*

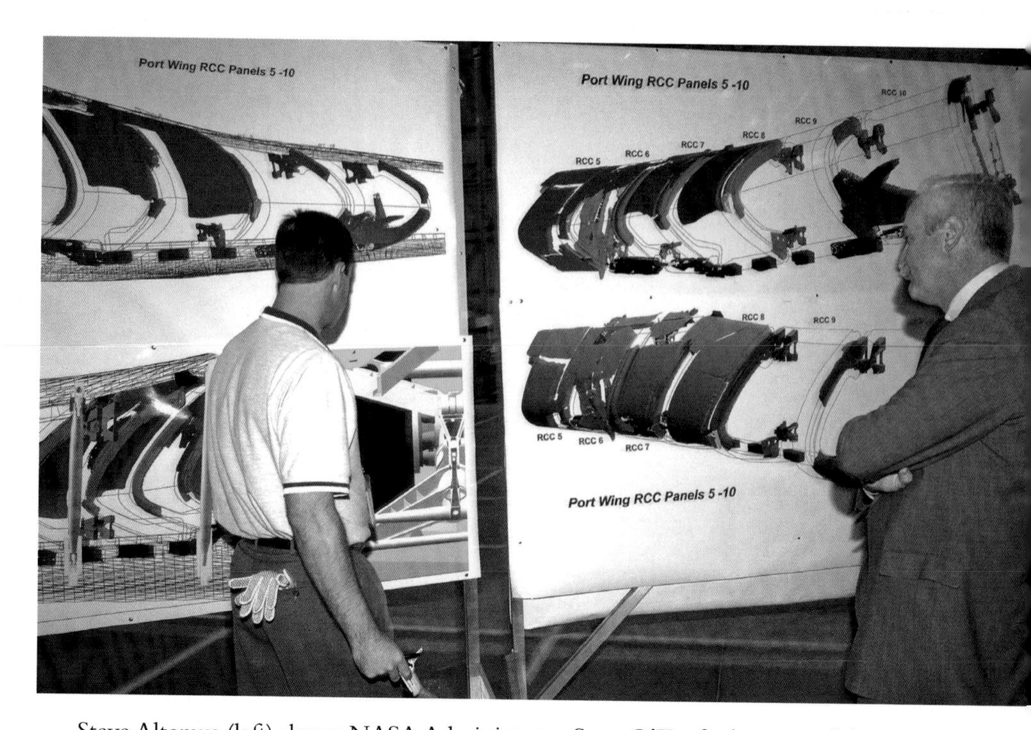


Steve Altemus (left) shows NASA Administrator Sean O'Keefe the state of the reconstructed leading edge of *Columbia*'s left wing as of April 28, 2003. The large amount of missing materia from RCC panels 8, 9, and 10 and their support structure provided overwhelming evidence th the wing was breached in that area. *(NASA photo)*

Ann Micklos's watch, with the time of the accident frozen on i face. *(Photo courtesy Ann Micklos)*

A cross, erected shortly after the accident, marks the spot where the first remains of a *Columbia* crew member were discovered near Hemphill, Texas. *(Jonathan Ward photo)*

Recovered sections of *Columbia*'s airlock and tunnel to Spacehab in the *Columbia* Preservation Office in the Vehicle Assembly Building. *(Robert Pearlman/collectSPACE)*

*Columbia*'s nose landing gear and some of the wrecked components from its avionics bay in the Columbia Preservation Office. *(NASA photo)*

Mike Leinbach and LeRoy Cain share a tearful hug at the Shuttle Landing Facility at the conclusion of the STS-135 mission and the end of the Space Shuttle Program on July 21, 2011. *(NASA photo)*

Hemphill's memorial star was transformed into a memorial to the sacrifices of *Columbia*'s crew and to Buzz Mier and Charles Krenek, with the new design unveiled on the sixth anniversary of the accident. *(Courtesy Patricia Huffman Smith "Remembering* Columbia*" Museum)*

The Patricia Huffman Smith "Remembering *Columbia*" Museum in Hemphill. Some of the exhibits include personal artifacts from the STS-107 crew, a space shuttle launch and entry suit, sample RCC panels, and a space shuttle cockpit simulator. *(Jonathan Ward photo)*

*Challenger* fuselage section and *Columbia* cockpit window frames in the "Forever Remembered" memorial in the Space Shuttle *Atlantis* building at the Kennedy Space Center Visitors Complex *(NASA photo)*

grieving. I did not want premature speculation circulating about the cause of the accident until more debris and more data came in. Finally, I did not want the important work of the engineers and technicians to be disrupted by having "outsiders" constantly looking over their shoulders.

After a few weeks, I felt it was appropriate to begin showing the outside world the work that was going on in the hangar. The first press visit consisted of my opening the door to the hangar and letting the press take photographs from that spot, but no closer. For the next several press opportunities, I took them a few feet into the hangar to a small area near the nose cap and let them look around from that one vantage point. Finally, by early April, we had a walkway around the hangar roped off, and I permitted the press to walk around. We never allowed them into the crew module area. I also escorted tours of congressmen, senators, governors, and other dignitaries throughout the reconstruction effort.

The sight of *Columbia*'s debris on the hangar floor and the smell of the charred materials never lost their emotional impact, especially for first-time visitors. Renée Ross came to the hangar to work with Pam Melroy in the crew module area. As they walked into the hangar on Ross's first visit, Melroy was making small talk and asking about her trip from Houston. Ross, though, could not hear what Melroy was saying to her. "I just stopped," Ross recalled. "The sight of the landing gear upside-down in the middle of the room, was just . . . I don't know how to explain it. It was breathtaking. I just stood there and looked at it all. I knew the vehicle's dry weight was 140,000 pounds. And there probably wasn't 20,000 pounds worth of stuff on that floor." Melroy kept talking, and then she suddenly realized that she was ten steps ahead of Ross. "She turned around to me and said, 'I'm sorry. I forgot that you haven't been here before. I need to step back and give you a few minutes.'"

No one was more surprised about the contents of the hangar than were the KSC workers who had been out in the field on recovery operations. The people in the East Texas collection centers had been compelled to get the recovered material to Barksdale, and then on to Kennedy, as quickly as possible, but they did not really comprehend what was happening with it after that. "We tried to ask for even one more day to do triage," Ed Mango said. "But the pressure was to get that stuff out to Barksdale! We couldn't understand why [the people in the hangar] needed it so fast."

Mike Ciannilli said, "We had no clue what was going on back at KSC. The first time I went to the hangar, I was blown out of the water. I was just mesmerized by it."

The sparse appearance of the shuttle on the grid gave the false impression that not much of the vehicle had been recovered. Only the items from the ship's external surfaces were placed on the grid, which were some 2,700 of the 84,000 pieces of debris recovered.

—

Engineers in the hangar began their detailed analyses of the debris to search for clues about the cause of the accident and how the ship came apart. As the NTSB predicted, the debris told a compelling story. Engineers just needed to sift through all the evidence.

Various parts of the vehicle showed signs of one or more significant abnormal conditions—extreme heat, oxidation, and mechanical stress. Some damage occurred before the ship broke up, and some afterward. The challenge would be to tease out the sequence of events and determine what had caused the ship to break up, as opposed to the damage suffered after the catastrophic event.

The effects of the great heat of reentry were visible everywhere. Some heat-resistant tiles were badly slumped—their shape distorted by partial melting—indicating they had experienced much higher heat than they were designed to withstand in a normal reentry. Globs of metal were spattered on interior and exterior surfaces. We found soot on pieces of the left wing's internal structure as well as on the left-hand OMS pod—a bulbous section under the shuttle's tail covering the maneuvering system engines. One of the "elevon actuators"—which controlled a steering flap on the left wing—was pierced by a sharply defined hole, as though someone had applied a blowtorch to it.[12] Aluminum globules were also found inside the crew module, on pieces such as seat belt fragments. Heat damage inside the crew module was almost certainly due to melting after the accident, as the vehicle broke up and the pieces reentered the atmosphere at high speed.

Oxidation of metallic objects made them appear to have been left rusting in a junkyard for years. Oxidation could also cause the reinforced carbon-carbon panels on the leading edge of the wings to erode and lose their effectiveness.

Everywhere were signs of the tremendous mechanical stresses to which the orbiter was subjected. After the "catastrophic event" in which the ship first broke up, *Columbia* was shredded into ever-smaller pieces as it encountered the denser air of the atmosphere at high speed. "Broomstraw fracturing" provided evidence of aluminum being twisted under high heat and stress—creating the appearance of the metal delaminating, which was a phenomenon new to many of the engineers.[13] Torsional forces exceeding 20 G had twisted and broken the seat frames in the crew compartment's mid-deck.

Engineers and materials scientists went to work using sophisticated equipment to examine the debris that appeared to be directly related to the accident as their top priority. At the same time, people who worked on various subsystems that were not directly related to the accident wanted to see how their components had performed.

Finally, there was the sensitive issue about discovering how the crew module had reacted to the breakup of the shuttle and what the crew might have experienced before and after the accident. This effort was technically not related to the determining the cause of the *Columbia* accident. Pam Melroy and her team made persuasive arguments about the value to be gained in studying how the various systems in the crew module either protected the crew or failed to help them during the aftermath of the accident. The crew module team conducted at least three audits during the reconstruction process, touching every piece of material in the crew module room, ensuring that the GPS and other identifying information were entered correctly in the database, and adding descriptive keywords that would facilitate future database searches.[14]

Incidentally, a few years after the accident investigation concluded, NASA was finishing up the process of addressing all of the observations and recommendations from the CAIB report. Melroy received permission to initiate a project to study the debris of *Columbia*'s crew module, which would close out one of the CAIB's observations. This study would provide detailed guidance for designing crew protection systems on future spacecraft.

To us in the reconstruction hangar, it often felt like the rest of NASA was focused on pursuing their own pet theories about what caused *Columbia* to disintegrate, rather than letting the debris tell its story through detailed examination.

Premature announcements to the press proved to be both unprofessional and embarrassing. A news release from the CAIB on March 18 said investigators believed that a carrier panel closeout tile might have fallen off of the shuttle's left wing in orbit, which would have allowed plasma into the leading edge of the wing.[15] Steve Altemus said, "The problem is that the CAIB announced it but never talked to us about it." After the story broke, NASA management called us to ask what had led to the CAIB's conclusion. Altemus replied, "I don't know, because the carrier panel is at my foot right now, so I know it didn't come off in orbit."

Some people said that the foam hitting the wing must have caused the accident. However, mission managers were reluctant to abandon the conclusion they reached during the flight—that *Columbia* could not have been mortally wounded by the foam strike on ascent. After the accident, they continued to insist that the foam could not possibly have caused fatal damage. Some people were convinced that one of the tires might have ruptured and blown open the wheel-well door. And others believed that *Columbia* had collided with something in orbit or during her descent.

The debris refuted many possible failure scenarios. One was that the left landing gear had accidentally deployed, rendering *Columbia* unstable during reentry. The condition of the left landing gear strut did not support that theory. The chrome on the upper surface of the strut would have melted away had the gear deployed and been exposed to the full force of the reentry environment. That chrome was relatively intact, so the gear could not have deployed. The outboard tires on the left side also showed much more evidence of heat damage than the inboard tires, pointing to a breach in the wing somewhere other than the wheel well.

Some tiles found early in the recovery period exhibited orange or brown streaks that appeared to be gouged into the tile surface. People speculated almost immediately about similarly colored insulating foam coming off the external tank and gouging the tiles. Or, perhaps, it was evidence of something else impacting *Columbia* during reentry. Analysis of the streaks in the reconstruction hangar instead showed that they were melted Inconel from the leading edge attachment fittings on the left wing. This was material from inside *Columbia* that had blown out through a hole in the wing and had been deposited onto and melted into the surface of the tiles.[16]

Another scenario that we needed to disprove was the potential involvement of terrorists. NASA management and the Department of Homeland Security considered this only a very remote possibility, but it had to be checked out. I did not want my team to be worried about possible terrorism, so I brought in the FBI undercover. The FBI special agents appeared just to be regular researchers looking at the materials. They swabbed various pieces and found no evidence of explosive residue anywhere on the vehicle.

I never told the team about their visit.

—

The Spacehab double module carried in *Columbia*'s cargo bay almost completely disintegrated during the accident. Spacehab would normally never be exposed to the heat of reentry, so it was relatively unshielded except for thermal blankets. Even though the shuttle's cargo bay doors protected it during the initial phases of reentry, the module was exposed to the full fury of heat and aerodynamic forces when *Columbia*'s main structure broke apart. As with the interior of the shuttle's payload bay, much of the module's aluminum structure and its insulation blankets were melted or consumed in the heat and friction of reentry. The only major parts of Spacehab's structure that survived the accident were two long Inconel rods—which were found almost completely intact—and pieces of the dense bulkheads of the module.[17]

Many of the science experiments were destroyed outright during *Columbia*'s disintegration and reentry, and others were badly damaged. Those that survived were mostly the ones that the crew removed from Spacehab and stowed in the orbiter's crew module lockers prior to reentry. Some searchers had found canisters or bits and pieces of experiments in the field. Pat Adkins, for example, found a bag of thick, creamy material with a plunger mechanism attached to the bag, which had been part of the "OSTEO" medical experiment to study bone cell growth. Electronic data for some experiments had been transmitted back to Earth during the mission, and a few data tapes partially survived the breakup.

One experiment caused a rare celebration when our workers examined it in the reconstruction hangar. Searchers found a thermos bottle-sized container from an experiment involving a colony of nematodes—small roundworms. The container had been inside one of the lockers in the crew module, so it was relatively well protected until the locker hit the ground. We were amazed to see *living* nematodes inside the container when it was opened in the reconstruction hangar. Nematodes have a short life span. Because this finding was several weeks after the accident, these were likely the descendants of the original animals in the experiment. Nonetheless, there was joy in the hangar at finding something alive—passengers of *Columbia* who survived the accident. "You look for the glimmer of hope where you can find it," said Spacehab's Marty McLellan.

A thorny legal issue arose regarding if and how material from scientific payloads in the orbiter should be returned to the researchers. Payloads and experiments on a shuttle mission technically belonged to the scientists—the principal investigators who designed and funded the experiments. However, NASA impounded all of the more than 2,200 pieces of recovered debris from *Columbia*'s experiments. NASA's primary concern was to prevent anyone from selling the recovered debris as memorabilia to recoup their lost investment in the experiment. NASA established a process to allow the payload customers to petition for access to the experiments to recover scientific data. Both NASA and the CAIB had to approve the request. Most of the science recovery operations took place at the reconstruction hangar.[18] The materials were then returned back to NASA custody.

All told, nine of the eighty experiments carried by *Columbia* were found inside metal boxes. Scientists who opened the containers believed that at least five of those experiments would yield usable data.[19]

—

Initially, the reconstruction team planned to lay out the pieces of wing tile on the grid on the hangar floor. Weeks into the process, engineers realized that having a few tiles laid out on the floor, still inside their collection bags, was not going to provide the big picture of how the left wing failed.

Each tile on *Columbia* was of a unique size and shape. We had all that information in our databases before the accident. Engineers tried using a modified version of KSC's EMAP software application—designed to track the status of waterproofing the shuttle's tiles in the Orbiter Processing Facility—to construct a virtual three-dimensional model of the recovered tiles and where they came from on the vehicle. While it showed which tiles had been recovered and which had not, it was not useful for seeing patterns in how the recovered tiles were damaged. We needed to use the tiles themselves to show us what happened.

In late March, we constructed a "tile table"—a platform with a full-size dimensional drawing of the 2,800 unique tiles on the left wing. As each piece of left wing tile came back, engineers identified it—sometimes based on a few millimeters of thickness—and placed it in the appropriate location on the tile table.

The tiles laid out on the table told a compelling story of the left wing's disintegration. The carrier panel tiles, which were the closeout between the panels on the leading edge of the wing and the tiles covering the rest of the wing, clearly showed where the wing breach occurred. The carrier panel tiles behind the inboard leading edge panels 1 through 8 appeared relatively similar. The tiles behind panel 9, though, showed evidence of high heating. Their surfaces were slumped, and their undersides were coated with metallic deposits from interior portions of the wing.

From the burn patterns on the other tiles, we saw that hot plasma had entered the wing at high velocity—thousands of miles per hour—and pressurized the wing cavity. The pressure created vents, which blew the superheated plasma and molten metal out of the upper and lower surfaces of the wing. The materials blowing out through the lower vent formed an obvious burn pattern along the underside of the wing.

As the plasma stream cut through the leading edge spar, it heated the wing and caused the adhesive that held the tiles onto the wing surface to fail. Those tiles peeled off the wing. They were designed to take heat from their outside surfaces, not from the side where they were glued onto corresponding felt pads covering the ship's wings. The degraded glue on the underside of those tiles clearly demonstrated that the wing was baking from the inside out.[20]

The heavily damaged tiles on the left OMS pod (at the left rear of the shuttle) and the left side of the vertical stabilizer also provided clear evidence that the insides of the left wing were melting. The airflow around the vehicle during reentry put the OMS pod and tail directly downstream of the wing. As the interior components of the left wing melted and burned, those materials were deposited on and heavily pitted the left OMS pod and left side of the tail. The right side of the tail and the right OMS pod exhibited none of that kind of damage. The right wing had therefore not melted prior to the orbiter's breakup.

Plotting the tile recovery locations on a map also supported the theory about the location of the breach in the wing. The tiles found farthest west in the debris field all came from the left wing, in the areas behind leading edge panels 8 and 9. Many of these tiles had brown streaks on them—Inconel metal from the melted leading edge attachment fittings.[21]

We also needed to find out what happened to the leading edge of *Columbia*'s left wing. Many pieces of the reinforced carbon-carbon (RCC) leading edge of the shuttle's wings came back from the field, in sizes ranging from larger than one square foot to smaller than a thumbnail. Boeing's Mike Gordon and NASA structures engineer Lyle Davis spent long hours using micrometers to study the thickness of each tiny piece. They determined which wing and which RCC panel they came from and how they fit together. It was painstaking work. Eventually, the leading edges of both wings began to take shape.

Since the leading edge panels are U-shaped in cross section and up to several feet long, it was difficult to fit them together meaningfully on a two-dimensional grid. First, the team tried tacking the pieces to Styrofoam shaped like the wing's leading edge. This proved unsatisfactory, because the backsides of the RCC panels were not visible. It was impossible to examine the panels in relation to the pieces of support structure that were behind them on the shuttle's wing. In April, KSC's shops fashioned three-dimensional frames made out of clear polycarbonate to hold the pieces of RCC and their supporting attachment fittings in their correct orientation. These frames enabled investigators to look at the reconstructed leading edge components from all sides.

As with the tile tables, the RCC panels told a compelling visual story about the accident. Panels 1 through 7, on the inboard side of the wing, were fractured and broken from forces after the shuttle disintegrated. Globs of aluminum and other metals were spattered along their inner surfaces. The metallic attachment fittings that held the RCC to the wing were still partially intact.

At panels 8 and 9, no metallic fittings were found. The support structure here was made of stainless steel, which melts at 2,500°F—a much higher temperature than the aluminum components in other areas of the wing. Heavy slag buildup inside the RCC panels implied that the leading edge spar behind those panels melted and then was deposited as molten metal onto the surfaces of the panels. The edges of the retrieved pieces of RCC were heavily eroded and knife-edged—signs that plasma acting as a blowtorch at over 3,000°F was applied at high pressure to the panels over a prolonged period of time. This was the only place on the wing where this pattern was observed.

From panels 10 outward, more of the metallic fittings were found, and there was less slag and no erosion.

Materials scientists analyzed cross sections of the slag deposits inside the left wing panels to determine what materials they were made of and how the deposits were laid down. There were several layers of material, which told the story of the wing's failure in time sequence. As more of the wing structure melted, the different types of metals from various parts of the wing were deposited in layers on top of the material already laid down on the inside of the panels.[22]

The location of the debris on the ground provided yet more evidence of how the shuttle came apart. Wreckage from the wings was not distributed randomly across East Texas. For example, the leading edge components from the middle to the tip of the left wing (panels 8 to 22) were found in the farthest west part of the debris field, between Dallas and Palestine. The leading edge pieces from the part of the left wing closest to the orbiter body (panels 1 to 7) came down between Palestine and Nacogdoches. The right wing leading edge pieces were the farthest east, scattered between Palestine and Hemphill. This provided evidence that the left wing failed before the right wing, the most likely failure point being near RCC panel 8 or 9.[23]

By late April, the story told by the debris was inescapable. The breach in the left wing was clearly in the wing's leading edge, at panel 8 or 9.

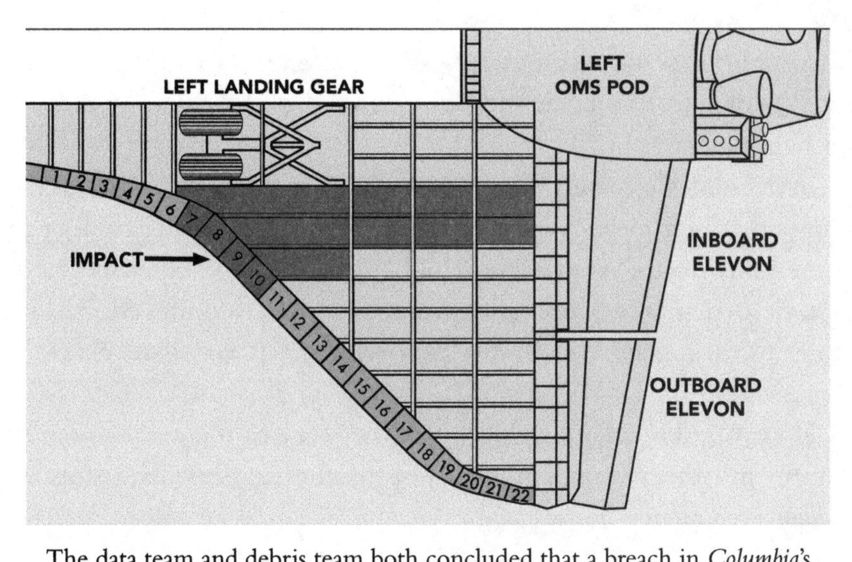


The data team and debris team both concluded that a breach in *Columbia*'s left wing near RCC panel 8 or 9 allowed plasma to enter the wing. Flowing at several thousand miles per hour and with a temperature well in excess of 3,000°F, the plasma acted like a blowtorch and melted much of the wing's support structure in the shaded area of this diagram.

On May 1, NASA announced that only one single failure scenario would explain all of the evidence contained in the debris, the OEX recorder, the telemetry received in Houston, and the videos provided by the public and other sources. Something—most likely the collision with the foam from the external tank—caused a breach in the leading edge of the left wing. This breach allowed plasma to penetrate the wing and erode it from the inside during the reentry. Whether the foam actually penetrated the leading edge—or if it pushed spacers apart and created a gap in the panels—will never be known.

The mystery of *Columbia*'s demise had been conclusively solved.

The last truckload of debris arrived at the reconstruction hangar on May 6. Wreckage that had been strewn over thousands of square miles of forest and field—pieces that had been carefully collected by tens of thousands of searchers working in tough conditions for three months—was all here in this one place. The twisted, burned, and shattered debris had told an important story, one that would have been impossible to write without the collective efforts of so many dedicated people in East Texas and Louisiana and of the nation's wildland firefighting teams.

In the end, *Columbia*'s debris represented hope for the future of the program. "Each piece was evidence of how hard *Columbia* fought to come home to us," Pam Melroy said. "We saw every recovered piece as a victory." Every piece of debris moved the reconstruction team closer to their goal: *We will find the problem, fix it, and move forward in their honor.*[24]

However, this was not the end of the story of the reconstruction hangar and the debris. This was just another intermediate stop on *Columbia*'s journey home.

# PART IV

# A BITTERSWEET VICTORY

We do not know where this journey will end, yet we know this: Human beings are headed into the cosmos.

*—President George W. Bush, January 14, 2004*

## Chapter 12

# HEALING AND CLOSURE

The long process of bringing closure to the story of *Columbia* began on February 11, 2003, when the final crew member was recovered in Sabine County. That same day, Ilan Ramon was buried at Moshav Nahalal in the Jezreel Valley in Israel. Rick Husband was buried in Amarillo, Texas, on February 21. A few days later, a memorial service was held for Kalpana Chawla at Zion National Monument in Utah, one of her favorite places. Willie McCool was buried in Anacortes, Washington, on March 4. Mike Anderson was buried on Friday, March 7, at Arlington National Cemetery. In a completely private gathering, President and Laura Bush spent nearly two hours with the families at the White House after Anderson's service. After they spent an hour in conversation in the Oval Office, the president personally took them on an impromptu hour-long tour of the residence.[1]

On what would have been her forty-second birthday, Laurel Clark was laid to rest at Arlington on March 10. Dave Brown was interred nearby on March 12. Clark, Anderson, and Brown's graves are near the *Columbia* Memorial and within a short distance of where the *Challenger* astronauts are interred. Dave Brown's brother Doug held a celebration of Dave's life after his interment. The family and friends tried to keep the mood light—an occasion to swap funny stories about Dave and the impact he had on their lives.

During the celebration, Doug took Ann Micklos aside and said, "I'm not supposed to tell you this, but—they found your watch." She was floored.

Because of the privacy surrounding the crew module reconstruction area, none of the workers in the rest of the hangar knew anything about the crew's personal effects that had been recovered. As she talked to Doug, Micklos learned that many personal items had already been returned to each of the crew's spouses. NASA told Doug about the watch. But because of Ann's "unofficial" relationship with Dave, she was not considered a family member. Protocol prevented NASA from releasing information about the recovered watch or returning it to her.

Doug fully supported her getting the watch back, and he helped start the paperwork to have it released to her. At the end of April, he called her to say that he had the watch in hand. She asked him to overnight it to her.

As she opened the package and inspected its contents, the condition of the watch amazed her. The leather band had burned away during reentry, and the crystal had shattered. However, the face of the watch was intact. The hands had stopped at 9:06—either when the crew module broke up or when the watch impacted the ground. Momentarily putting aside the sobering thoughts about what the watch had been through, Micklos laughed to herself that Dave had been thoughtful enough to set the watch to Eastern Time for her.

Micklos took the watch back to the reconstruction hangar to show Pam Melroy's team she had received it. They were ecstatic to finally be able to discuss it with her. The watch had been processed through the crew module room, where it was treated like every other item of crew personal effects. The team had known for weeks it was hers, but they observed protocol and did not say anything to her about it. With the veil of secrecy finally lifted, everyone could celebrate the watch's recovery and return.

Melroy told Micklos several other watches had also been recovered. Micklos's watch, however, was the only one with the time of the accident preserved on its face. The crew module team had even debated as to whether that time indication could be used as evidence in the investigation. However, they concluded that because the watch was considered personal property and not official flight hardware, it could not be used for evidence.[2]

The story of the watch has one final chapter. On February 1, 2004—the first anniversary of the accident—Micklos asked to go out to the Shuttle Landing Facility runway with a couple of her colleagues to reflect on the loss of *Columbia.* USA shuttle landing technician Billy McClure volunteered to drive them.

Micklos went out onto the runway and had a moment of peace despite the morning's heavy rain. Once back in the van, she and the group exchanged stories about their experiences with *Columbia.* McClure said he was one of the first responders from KSC to collect debris in Texas. He recalled walking along, scanning the ground, when something shiny caught his eye. He bent down to look at it, and saw it was a rectangular watch.

Micklos asked, "Did it have a blue face?" He said it did. Micklos was stunned. Out of all the people who searched for debris from *Columbia,* what were the odds that she would be in a van on the anniversary of the accident with the man who had found her watch? The astonishing coincidence brought everyone in the van to tears.

—

During the course of the spring and summer, large and small acts helped the recovery and reconstruction teams begin their own healing process. For many people, the most difficult thing they faced turned out not to be the grueling days in East Texas or long days on the hangar floor. It would be much harder for them to cope with the challenges they faced in returning to "normal" life, when the enormity of events of the winter and spring began to sink in.

KSC's staff had to keep their emotions in check during their assignments in the field in order to cope with the intensity of the work situation. Pat Adkins said, "I didn't have time to dwell on the fate of the ship and crew. I just lost myself in the work at hand, and I consider that a weak balm for it all."

"It was pretty traumatic," said René Arriëns. "I couldn't even describe it to my wife. Only the people who worked with you could have understood it. I know Jerry Ross did. I could see it in his eyes. I saw him once in the command post. I just looked at him, and I knew."

Many people who were required to rotate out before the recovery operation ended were upset about leaving the work unfinished. Greg Cohrs said, "I felt such a tremendous, intense sense of obligation to complete this mission that I requested—nearly begged—to be able to stay on."

USA's Linda Moynihan said, "It was like when a soldier's time is up and he leaves his buddies there in the war zone. You feel like you left your friends there and your responsibilities behind."

Brother Fred Raney occasionally counseled NASA workers—people who were so driven to perform meaningful work in service of *Columbia* and her crew that they could not bear to leave before the task was completed. Raney said, "I talked to one man who was literally in tears because he had used all of his vacation and sick time, and he had to go back to his job in Houston."

The astronauts who participated in crew remains recovery were required to meet with a counselor at the end of their time in the field.[3] Astronaut Dom Gorie said, "We're often accused, as test pilots and military people, of being over-compartmentalizers. We compartmentalize so well that some people can't really see the human element behind things like this. And I've often been asked, 'How could you fly the next day, after you lose an F-18 in the Gulf War?' It's because you have to compartmentalize. We had to do that. But it's pretty tough when you get asked to go out and pick up remains [of] your friends, or from the shuttle that was carrying them. And the way to get through that was trying to stay focused on what the endgame was."

"We all knew *Columbia*'s crew," Gerry Schumann said. "We knew them by their first names. We knew their wives, their husbands, and their families. That was the toughest part—recovering remains of a crew that you knew." Schumann lost himself in his work, pulling seven-day weeks for several months. "It was easier for me to continue working than to take any time off and think about everything," he said. "One day they made me go buy a fishing pole, and dropped me off at a lake to go fishing. I sat there for about half an hour and said, 'Okay, I'm ready to go back.'"

Counselors visited the US Forest Service fire crews at their camps to talk to searchers who needed help dealing with the nature of the accident, the intensity of the work, and being so far from home. At one morning's briefing in the Hemphill area, one of the Forest Service leaders introduced the counselors to the broader leadership team. He then turned to Schumann and said, "Gerry, they're here to see you."

Schumann was irate. He refused to speak with them or even acknowledge their presence for the week they were on-site. "The big joke was—when we left Texas, the team gave me a stack of a hundred business cards they'd collected of counselors I could talk to," Schumann said.

After the camps shut down, the last KSC contingents from East Texas and Barksdale returned home on two chartered planes out of Shreveport, one led by Ed Mango and one by Jeff Angermeier. Mango thanked everyone for their sacrifices. Then he cautioned them to remember that their families had not shared the experience. Their families would not be able to fully comprehend the intensity of what the searchers had endured. Mango reminded his group, "Our families all had normal lives. *Our* lives are the ones that really changed. We are the ones who are going to look at life differently now when we get back. We can't expect them to understand." He also informed the group about some of the statistics regarding the results of the recovery operations, so that the people could start to get a feel for the overall size of the effort of which they had been a part. For some people, it was the first time they had heard the magnitude of the accomplishment.

The workers' families greeted everyone on their arrival at Kennedy's runway. Mango likened it to a homecoming after an overseas combat deployment. The atmosphere was festive. Several astronauts with their T-38 jets were on hand to thank the families and welcome the staff home again.

The shock of immersion back into "normal life" and the less hectic work pace allowed months of pent-up feelings to surface. Grief, guilt, and anger were the predominant emotions. "The toughest time for me was the first week of May after we came home," René Arriëns said. "You just kind of started to take in and absorb everything that happened over the last three months. I spent about a week being an emotional wreck."

Gerry Schumann's wife Gail said, "He was angry when he came home. He was just not the same person that went. He was very, very angry for many years."

Gerry said that when he went back to his office the day after returning from Texas, his boss was ". . . sitting there with his feet up on his desk. I lost it. I cussed him out. Then I left for two weeks and didn't ever want to come back."

Schumann was ordered to attend counseling, as were many other people who returned from the field still traumatized by what they had experienced. It took four months before he was even able to sit down with his wife to talk about what he was feeling. His anger was focused on his boss, himself, and then with the whole NASA system—for failing to speak up or do something to prevent the accident. "We were focused on the only job we had, which was to make sure that the vehicle was safe to fly, and if something else happened out there with the operation itself, that wasn't our problem," Schumann said. "When you look at it in hindsight, what would it have taken for us to say something?"

People berated themselves for real or perceived missed opportunities to have spoken up, asked difficult questions, or done something differently. Whether it would have changed anything is another matter. The CAIB's report on the accident noted that NASA's culture made it extremely difficult to raise concerns that would have been listened to. That was clearly the case in the questions raised during the Mission Management Team discussions about the foam strike.

I know from personal experience that it took big balls to bring something up at any of those meetings, even if you damn sure knew you were right.

Second-guessing one's actions was not limited to the NASA family. Jeff Williams, who generated maps for the search effort, said, "I'm not the only one who could only sleep three hours per night. I woke up and started thinking about all the things I could have done, should have done, would have done."[4]

The trauma of being involved in recovering the crew's remains haunted the volunteers and the searchers in the US Forest Service for months afterward. "After it was over, the astronauts came to town and wanted to visit with people. I didn't want to talk to them," said Marsha Cooper. "There was something I was dealing with, and I just didn't want to get close to them. I would have lost it. I wasn't ready. Felix Holmes and I talked a lot after it was over. He called me one day about a month or so after it was all over, as he was just going down the road. He had finally lost it. We all had a breaking point. We just didn't know when it would come."

—

On March 31, Kirstie McCool Chadwick, sister of *Columbia* pilot Willie McCool, and I were special guests at the Florida Marlins home opener game in Miami. Huge *Challenger* and *Columbia* plaques decorated the right center-field wall. We came walking out from behind one of the outfield walls and received a standing ovation. People were hollering down from the stands, "Thank you!" It was great to see that the folks in Miami recognized what was going on up in Brevard County.

Closer to home, KSC held several events during May and June to thank Kennedy's workers and their families for their contributions and sacrifices in the recovery and reconstruction efforts.

On May 7, our reconstruction hangar team and their families were the guests of honor at a Florida Manatees minor league baseball game at Space Coast Stadium in Viera. It was important to give our workers some relief. We all needed to blow off some steam—to party and forget about the hangar a little bit. We tailgated in the parking lot before the game. Steve Altemus and his team leaders boiled eight hundred bratwursts in beer and grilled them with red onions. I threw out the game's opening pitch, which promptly landed in the dirt.

At the end of the evening, our group persuaded Ann Micklos to dance on top of the dugout. Micklos later explained, "To me, this was all part of the grieving process. You needed that big family to get you through this. You've got to make a joke when it gets too serious. That's what got us through—that balance between the two. Because if you didn't have that balance, we never could have done what we did."

We held a daylong picnic for KSC's *Columbia* teams and their families at KARS Park near Kennedy Space Center on May 30. It was the first time the recovery teams from all of the sites came together after the accident, and it was the first joint event ever held for both the recovery and reconstruction efforts. The party was a blowout of mammoth proportions. Hundreds of workers and their families attended. Many stayed overnight in RVs. Jim Comer and his helpers shucked 120 cases of corn and cooked 100 pounds of shrimp. Our finest rocket engineers designed and built an "atomic slip-and-slide" for the kids, and many adults found their way onto it as well. Water-gun fights and volleyball games helped people loosen up and celebrate their hard work and friendships.

—

The exclamation point that provided closure to the accident investigation was independent of the reconstruction and data teams. NASA and the CAIB wanted to simulate as accurately as possible the launch debris strike on *Columbia*. To do this, they needed to shoot foam insulation into samples of the shuttle's silica tile and reinforced carbon-carbon wing panels at the speed and angles at which the foam would have collided with the ship during the impact.

Two days after the accident, the CAIB and NASA contacted the Southwest Research Institute (SwRI) near San Antonio for assistance in the accident investigation. SwRI had conducted previous studies for NASA on the effects of impacts of much smaller pieces of foam, cork insulation, and ice on shuttle tiles. The Institute was a logical choice for performing the tests that would simulate the conditions of the *Columbia* accident.

Investigators wanted incontrovertible proof that foam from the external tank was capable of inflicting mortal damage on the shuttle's thermal protection system. That foam could damage the wing seemed counterintuitive on many levels. *How could a piece of lightweight insulation—about the density of Styrofoam and weighing less than two pounds—fall off the tank and cause that kind of damage? And wasn't it traveling at about the same speed as the shuttle?*

In fact, analysis showed a significant velocity difference between the shuttle and the foam at the time of impact. NASA estimated that the shuttle was traveling faster than 1,500 mph—and accelerating—when the foam fell off the tank. After falling off, the foam immediately and rapidly decelerated due to air resistance. The block slowed to about 1,000 mph in the 0.2 seconds between when it came off the tank and when the shuttle's wing impacted the foam. The relative difference in speeds between the shuttle and foam was therefore more than 500 mph.[5]

The piece of foam that struck *Columbia* was four hundred times larger than the pieces tested previously by SwRI. Using a special compressed air cannon, SwRI planned to simulate the collision by firing foam blocks at more than 500 mph into samples of shuttle tiles and wing leading edge panels. High-speed cameras photographed the test firings and impacts, and over two hundred sensors measured the effects of the collisions.

By the time the equipment and procedures were ready for the first test on the landing gear door, the investigation had already narrowed its focus to the wing's leading edge as the impact area. SwRI ran its test anyway using a landing gear door—one borrowed from *Enterprise*[6] and subsequently covered with silica tiles—to check out the test equipment and processes. As expected, a grazing impact of foam, akin to what would have occurred in flight had the foam hit the underside of the wing, caused only minor damage to the tiles on the landing gear door.

Space shuttle wing leading edge panels are large, expensive, and made to order. The reinforced carbon-carbon (RCC) material also wears and becomes more brittle over time, so SwRI could not use newly manufactured panels to get an accurate assessment of potential damage in its impact tests. The test panels would have to come from the wings of *Discovery* and *Atlantis*, the two orbiters that had flown about as many times as *Columbia*. NASA decided to test the process first using fiberglass wing panels from *Enterprise*, which was not designed to fly in space.

Several test shots at *Enterprise*'s fiberglass panels—which were stronger than the RCC panels on the flight-worthy shuttles—produced scuff marks from the foam blocks, but no breakage. After getting its process and equipment calibrated, SwRI was now ready to try the tests with the space-flown RCC panels.

First, a foam block was fired at panel 6 from *Discovery*. The impact created a crack nearly six inches long in a rib supporting the leading edge, and it moved the panel enough to create a small gap in the T-seal between panels 6 and 7. This test proved that foam could damage the RCC material. However, the damage incurred in this test would not have been severe enough to create the burn-through seen on *Columbia*. NASA estimated a hole of at least ten inches in diameter would have been needed for the wing to ingest a plasma stream large enough to create the damage seen in *Columbia*'s debris.

The next test target was panel 8, which had flown twenty-six times on *Atlantis*. Evidence from the reconstructed debris and the OEX recorder indicated that panel 8 was the site of the impact on *Columbia*'s wing.

At the test on Monday, July 7, the impact from the foam block blew a huge hole through the panel about sixteen inches by sixteen inches across, created several other cracks, and caused the T-seal to fail between panels 8 and 9. This was entirely consistent with the type of damage postulated to have caused *Columbia*'s demise.

Witnesses were incredulous, but the evidence was incontrovertible. NASA now had the smoking gun matching the fatal wound on *Columbia*. The test silenced lingering doubts that a foam strike alone was sufficient to damage the wing and doom the ship.

—

In the late spring, we invited the families of the STS-107 crew to visit the reconstruction hangar and see *Columbia*'s debris. Our staff prepared carefully for the visit. We wanted everything to be as perfect as possible for the families.[7]

We briefed the spouses in advance to prepare them for what they would see. To make the atmosphere as private as possible, only a few of the staff were on hand when the spouses arrived with their astronaut escorts. I greeted them at the entrance, and Steve Altemus escorted them through the hangar.

Lani McCool was interested in *Columbia*'s cockpit window frames. She asked me about the windows and where her husband Willie was sitting during reentry. I told her, "As the pilot, he would have been sitting on the ship's right side."

She asked, "So, he was behind these windows?"

I said, "Yes, that's where he would have been." She then reached into her bag, pulled out a flower lei, and placed it behind a window.

Then it got tough. She asked me point-blank, "What do you think Willie knew?"

I said, "I don't know. Lani—your husband was one of the best pilots in the world. He knew if he had an aircraft that was in control or not. I'm sure that at some point, he knew he was in trouble." She thanked me for being straight with her.

The spouses returned a few weeks later for another visit, this time with some of their children.[8] The children were at first reluctant to enter and had to be encouraged to come into the hangar. John Biegert and Robert Hanley escorted them through the crew module area. "For me, it was the most emotional time in the whole process," Biegert said. "The kids just wanted to hold or touch something that their mom or dad had touched or had been near. We had built up the seats as best we could, and had the flight equipment and hand controllers and switch panels. It was hard for them to see that, but they felt that they had to do it. Some of the kids came back five years later, and I took them through the crew module again."

The families appreciated being able to see what had happened to the orbiter. They also witnessed the reverence and care with which we were treating the vestiges of *Columbia*—and the lengths to which NASA had gone to understand and learn from the accident—to ensure that future space travel would be safer because of the sacrifices of *Columbia*'s crew.

In July, we opened the reconstruction hangar to all KSC workers and their families. We saw it as an opportunity "to educate and inform, give us a new respect for space exploration and those who serve, and allow closure to this tragedy in some ways."[9] We also knew it was important for the people who had been deployed to the recovery operations in Texas to see the results of their labors.

Many workers at KSC felt a sense of guilt over the loss of the shuttle, perhaps feeling they had somehow contributed to the accident or failed to prevent it. Emotions were still raw. Even the women who sewed the quilted fabric insulation blankets for the shuttles thought they might have caused the accident. "I had to personally go out there and let them know that they were not responsible for this loss. They were really upset about it," said Roy Bridges, who was KSC center director at the time.

We set aside Monday through Wednesday, July 7 through 9, for KSC personnel to visit the hangar. Then KSC employees and their families were permitted to visit between three and eight o'clock on

Thursday and Friday, and all day Saturday. I sat at the entrance to the hangar and personally greeted every visitor—eleven thousand people by the end of the week. It was tough going for an introvert like me, but I felt they needed to see a familiar face when they arrived in that unsettling environment. NASA, Boeing, and United Space Alliance employees were on hand to answer questions about the debris on the hangar floor, the test processes, and our conclusions about the accident. The *Sixteen Minutes from Home* video commemorating *Columbia* and her crew played at a screen off to one side of the hangar. The video never failed to touch people's hearts.

We stationed several employee assistance counselors throughout the hangar to help folks who were emotionally overwhelmed by the sight of *Columbia*'s wreckage. And indeed, a few people became distraught when the sight of the debris proved too much for them to bear. Even my wife Charlotte was startled at the powerful impact of the remains of the ship. She said, "I knew what the windows looked like, and I knew they were around the corner, but you just can't imagine it until you see it. You think you can imagine it, but you can't."

It was a remarkable opportunity for the greater KSC family to appreciate what they had accomplished in months of sacrifice and hardship, and to welcome their old friend *Columbia* back home one last time.

On Sunday, July 10, we conducted a private tour for the families of the women and men who had been working in the hangar for the past five months. Steve Altemus said, "The wives and husbands and sons and daughters of whoever was working in the hangar now suddenly got a sense of what their mom or dad was working on for so many hours for four months. Here's what they were doing and why it was important."

In contrast to the rapid, nearly instantaneous start-up of activity in the hangar in February, the reconstruction effort tailed off gradually during the early summer. Groups closed out their areas and moved on.

I gathered with some of my closest associates on the management team on the runway apron outside the hangar entrance near the end of July. I made a special request to the Center director for our team to have a couple of beers that evening. That was unheard of. You just didn't bring alcohol to "officially" drink on KSC grounds. We swapped stories, pined away at the thought of this phase of our association nearing an end, and toasted one another using space shuttle shot glasses that Jon Cowart supplied for the occasion. Then we drank a toast to *Columbia* and her crew, and poured a bottle of Scotch onto the runway.

On August 12, Roger and Belinda Gay, accompanied by their son, daughter, and several other relatives and friends, came from Hemphill to Kennedy Space Center as guests of NASA to tour the reconstruction hangar and see the shuttle processing facilities. The Gays spoke to workers at KSC, relating their account of the amazing events of early February and speaking of the bond that quickly formed between the citizens of Hemphill and the NASA family.

The Gays said that the Sabine County community was discussing plans to build a memorial to *Columbia* and her crew, the helicopter crew who perished during the search, and the community's volunteer effort. Texas A&M University architecture students were already developing preliminary designs for consideration.

Visiting the reconstruction hangar brought the whole recovery experience full circle for the Gays. Belinda said, "We needed to come here. Seeing the hangar was a very emotional experience and gave us some sense of closure."[10]

—

Our workforce had endured an incredibly grueling first half of the year. Between the loss of the shuttle and its crew, the hardship of the recovery operations, and the reconstruction, our staff was physically exhausted and emotionally drained. No one knew if or when the space shuttle would fly again. Contractors worried about their job security if the program were to be canceled or scaled back.

By the end of April, I realized our launch team needed a goal in order to push themselves through the grieving process. It was time to get the team back together, operating effectively in familiar surroundings, and focused on flying America's shuttles again.

I asked my managers to start putting together an intense schedule of launch countdown simulations. We would normally run these "sims" once per mission. However, since the program was on indefinite hold, there might be months or years of stand-down time before the shuttle was ready to fly again. I scheduled the first countdown sim for June 1, exactly four months after the accident. I wanted our team to run a full launch countdown sim every six weeks until the program was back on its feet. It would keep us sharp and future-focused, feeling good about our team, and ready to launch shuttles again.

Overcoming inertia to run sims again was challenging. The team had to take on a massive amount of work. However, people soon found themselves deep into the hardware and procedures again, pulling schematics and drawings, and working through problems. They gradually began to function as a cohesive team again.

The hard work was therapeutic. Some people thought I was pushing too hard, but responding to a tough challenge was what the team did best, and it was what they needed. It helped the team turn the corner from despair over the loss of *Columbia* to hope for the future of the program.

We didn't know it then, but it would be another two years before the space shuttle would return to flight. The book was not yet closed on *Columbia*, and the rules for the future of the program had yet to be written. However, we were finally able to start focusing on the future and putting the past behind us.

Chapter 13

# PRESERVING AND LEARNING FROM *COLUMBIA*

Ron Phelps was NASA's project manager for reconstructing *Challenger* after the 1986 accident. When the investigation concluded and it was time to dispose of the debris, Phelps and his team evaluated several sites at KSC and Cape Canaveral Air Force Station as storage facilities. They selected the Launch Complex 31 and 32 Minuteman missile silos at Cape Canaveral—each seventy-eight feet deep and twelve feet in diameter—as an appropriate location. Smaller pieces of *Challenger*'s debris were cataloged, packed into 102 crates, and stored in the silos' underground equipment and battery rooms.[1] Larger pieces of debris were lowered directly into the silo. One segment of a *Challenger* wing was too large to fit and had to be cut in half. Pieces of *Challenger* debris that still occasionally wash up on the Florida shore are placed in the side chambers.

The silos are sealed off with concrete caps, to be opened only under extraordinary circumstances. Although Complex 31/32 is officially classified as a storage site, most people consider it a burial site. Sadly, not even a marker or sign identifies the site.

NASA occasionally studied some of *Challenger*'s debris during investigations of specific issues on the other orbiters. For example, when we discovered cracks in the fuel system flowliners of several shuttles in 2002, our engineers examined the recovered flowliner assemblies from *Challenger* to see if the problem existed in 1986. Otherwise, *Challenger* has remained undisturbed and out of sight.

About one month into the *Columbia* reconstruction effort, I discussed with my team what to do with *Columbia*'s wreckage once the investigation concluded. I met with Phelps, who advocated burying *Columbia*'s debris as the easiest thing to do. A wave of emotion immediately overcame me. It is hard to put that feeling into words, but I knew at that moment that we had to do what was *right,* not necessarily what was easy. I came back and told my team, "I don't know what we'll do yet, but we are *not* going to bury *Columbia*." They were relieved. We couldn't just put her in the ground and pretend the accident didn't happen.

We discussed alternatives for the next several days. We knew *Columbia*'s collective debris had the distinction of being the most material ever to have survived a hypersonic breakup and reentry from the boundary between space and the upper atmosphere. The collected eighty-five thousand pounds of debris was ten times the amount recovered from all previous uncontrolled reentries combined.[2] Our materials scientists said *Columbia* was providing them a diversity of material and accident conditions that they had never studied before.[3] *Columbia* was hard evidence of the effects of high heat, aerodynamic stresses, and an ionizing oxygen plasma environment on a wide variety of materials used in spacecraft. NASA clearly needed to seize the opportunity to learn from *Columbia* to make future space vehicles safer.

I informally floated the idea of preserving *Columbia* past Sean O'Keefe, when the administrator was visiting KSC. He encouraged me to develop a plan. Sean's leadership was absolutely the key to the success of the entire effort. He set the tone for the agency from the outset—one of openness and a willingness to bring something positive out of the tragedy. He embraced the CAIB and its recommendations, and made sure everyone else did as well. This was so different from the mood of NASA following *Challenger*. Back then it was "put it behind us and move on."

I asked our vehicle flow manager Scott Thurston to lead a *Columbia* Preservation Team and develop ideas for storing and studying

*Columbia*'s debris. On May 9, 2003, the team issued a formal Request for Information to gather ideas from the scientific, academic, and government communities. We received fifteen letters of interest in June.[4] In addition to suggestions for storing and curating the material, the responses included recommendations for studying pieces of debris, using the material to teach failure analysis techniques, examining how various types of welded or bonded connections performed during *Columbia*'s breakup and reentry, and analyzing the trajectory of the debris as the vehicle broke up.

The enthusiastic reaction of the research organizations was heartening. The responses validated our vision of *Columbia* performing an important, ongoing scientific mission.

With approval to preserve *Columbia*, Thurston's team next examined options for an appropriate storage facility. Putting *Columbia* into a silo at the Cape would have challenged the objective of making the material accessible to researchers. They also considered storing the debris in a section of the reconstruction hangar, at the Spacecraft Assembly and Encapsulation Facility in KSC's industrial area, or in a leased facility near KSC.

By late July, the team identified a 6,800 square foot room on the sixteenth floor of A Tower in the Vehicle Assembly Building. The VAB afforded environmentally controlled space and secure but relatively convenient access to researchers. Since NASA already owned the facility, the costs of revitalizing the site and moving the debris would be minimal—only about $130,000.[5]

Pam Melroy and Jim Comer flew to Washington, DC, to meet with the families of *Columbia*'s crew and obtain their blessing for exhibiting the material for research and education purposes in the VAB. The families endorsed the proposal. They felt that the crew would have wanted *Columbia* to be used as a means of advancing scientific knowledge.

Once NASA leadership approved the VAB site, we started packing everything in the reconstruction hangar and began preparing the VAB room to receive the debris. Amy Mangiacapra, James Harrison, and Jack Nowling of United Space Alliance had all worked in the reconstruction hangar. Now, they would oversee the task of moving the eighty-four thousand pieces of *Columbia*'s debris to the VAB.

Most of the material went to the VAB the same way that it arrived from Barksdale—in bags inside large triple-wall boxes. The main difference was that the materials were now identified, inventoried, and sorted into tote trays inside the boxes. They were keyed to a database that made it easy to locate and retrieve any piece requested by researchers for study. As in the reconstruction hangar—but on a much smaller scale—the majority of the material would be stored off to the side in the boxes, with selected pieces on display on the floor or on racks.

Beginning on September 15, 2003, workers started loading all of the *Columbia* material onto flatbed trailers and transporting it one and one-half miles south to the VAB. Piece by piece, box by box, propellant tank by landing gear strut by RCC frame, a team of about forty workers took everything up the VAB elevator to the sixteenth floor and loaded it carefully into the preservation office. The move was finished by the end of September.

The reconstruction phase of *Columbia*'s story came to a quiet close.

Even though the recovery and reconstruction were officially over, residents of East Texas continued to find pieces of the shuttle that were overlooked during the ground and air searches. The great majority of the calls came from well-intentioned citizens who wanted to do their part to preserve *Columbia*. Jim Comer said, "One of the toughest emotional moments for me was when a lady sent in a Styrofoam plate with a letter that said, 'Dear Space Shuttle Team: I found this piece of foam tile in my back yard. God bless all of you for your work.' I broke down and started crying. I wrote her a letter back and thanked her."

The volume of calls tailed off after the first several years, but people still continue to find shuttle debris on occasion. One of *Columbia*'s liquid oxygen tanks—somehow missed during the navy's search in 2003—was exposed at the bottom of Lake Nacogdoches in late July 2011, when a severe drought caused the lake's water level to drop about eleven feet.[6] NASA and local officials retrieved the tank and took it to KSC to join the rest of *Columbia*'s debris.

Pieces of *Columbia* will probably continue to turn up for years. "There's still stuff out there," said Greg Cohrs. "The ground was really wet at the time of the accident, so there's stuff buried. It's part of the archaeological record now." I agree. We only found three of the six turbopumps. They're so massive; three more must be out there somewhere. Perhaps one of them was the car-sized object that people reported hearing splash into the Toledo Bend Reservoir.

Toward the end of search operations in the spring of 2003, NASA established the *Columbia* Recovery Office to handle calls from citizens about debris findings. Five employees from Johnson Space Center staffed the office until the function was transitioned to Kennedy in October 2003. This administratively consolidated the storage and coordination of *Columbia*'s debris in one location.[7]

The anniversaries of all three of NASA's fatal spacecraft accidents fall within a one-week period between January 27 and February 1. Sean O'Keefe designated the last Thursday in January as an annual NASA Remembrance Day for the crews of Apollo 1, *Challenger*, and *Columbia*. On that first Remembrance Day—January 29, 2004—KSC Director Jim Kennedy and I officially dedicated the *Columbia* Research and Preservation Office in the VAB.

Pam Melroy said at the ceremony, "I realized this facility is *Columbia*'s 'Arlington.' We have a very special place to come and reflect and be inspired. If you've ever been to Arlington [National Cemetery], it's a very inspiring place to see all the people that have sacrificed everything for the sake of our country. And it's the same thing here for the thousands of lives that built *Columbia*, maintained her, launched her, and flew her. That is our dedication here."[8]

—

Even before we officially dedicated the *Columbia* office, its new scientific and educational mission began. In November 2003, two teams from Johnson Space Center requested samples of tile and of *Columbia*'s right wing leading edge for study. The teams were tasked with developing methods and materials that astronauts could use to repair the shuttle's thermal protection system in orbit. Even had *Columbia*'s crew known about the breach in the left wing, there was probably nothing they could have done about it, since the shuttle did not carry a repair kit. The techniques that the JSC teams developed as a result of studying *Columbia*'s materials were subsequently tested in space aboard *Discovery* on STS-121 in July 2006.[9] In another study, NASA sampled some of *Columbia*'s high-pressure storage tanks to help certify that the other shuttles were safe to fly.[10]

One of the first academic uses of samples of *Columbia* debris was in a masters-level class in materials sciences forensics at Lehigh University. Jim Comer and several others from the reconstruction effort attended the presentations given by the students. Our people were impressed at the insights and conclusions the students garnered from the debris.[11]

Institutions, spacecraft designers, and other researchers can petition to borrow material from *Columbia* for study. Scott Thurston said, "*Columbia*, at her core, was a scientific vessel, flying highly scientific missions. Her ongoing contributions to science are the legacy of her and her crew. We would want them to know that they were still contributing to that mission."

Unlike a military crash investigation, the CAIB's analysis of the *Columbia* accident did not include a detailed survivability analysis of what happened to the crew and their equipment. Pam Melroy vigorously advocated conducting such a study during the accident investigation. However, it was considered to be outside of the CAIB's charter, and no funding was available to keep her crew module reconstruction team together after the CAIB confirmed the cause of the accident. Much to her relief, the Space Shuttle Program Office later agreed that studying *Columbia*'s debris offered a unique learning opportunity that could benefit future spacecraft design. In July 2004, NASA formed a multidisciplinary Spacecraft Crew Survival Integrated Investigation Team, with Melroy as one of the deputy project managers.[12]

The team visited KSC several times and conducted two major debris reviews using material pulled from storage for study. Melroy said she was particularly grateful for the extra effort expended by the crew module reconstruction team in documenting the debris in 2003, because it facilitated retrieving materials for this study.

Melroy recalled the most personally impactful moment of her study to be examining the ship's R2 control panel, which was immediately to the right of pilot Willie McCool in the cockpit. The switches on other control panels recovered from *Columbia* might have been jostled during the breakup, impact, or subsequent ground handling. Since the investigators could not be sure how those switches were configured just prior to the vehicle breaking up, the switch positions on those panels were not valid evidence for the investigation. However, the R2 panel was discovered in the field bent nearly in half, which protected the switches from being accidentally moved during the recovery. Some of the switches on the panel are lever-locked, requiring two actions to move them to a new position. When the investigation team pried the R2 panel open, they discovered that two of three lever-locked switches in one cluster were in a different setting than the third. These were the controls to cool down the shuttle's auxiliary power units before restarting them.

"As a shuttle pilot, I knew exactly what those switches meant," Melroy said, "and I was absolutely electrified!" The crew—"demonstrating excellent systems knowledge by taking actions not simulated in training"—had attempted to get the hydraulic systems working again after the ship lost control. Melroy said, "They were our heroes. Somehow, it made it easier to bear, knowing they were not helpless, but instead were fighting to save the vehicle in their last moments—right up to the end."[13]

The four hundred–page crew survival investigation report, released in 2008, confirmed that the accident was not survivable. Rather than the CAIB's general finding that plasma intrusion caused the left wing to fail, the survivability study presented a detailed analysis and timeline of how *Columbia* actually broke up. The proximal cause of the accident was the loss of hydraulic pressure after the system was breached by plasma in the left wing. This caused the ship's control surfaces to stop responding to steering commands. *Columbia* went into a flat spin. The crew knew their ship was in trouble. They tried to save it after it went out of control, during a period that lasted at most thirty seconds. The ship broke up due to aerodynamic forces, starting with the left wing. A breach in the crew module caused it to depressurize rapidly, shortly after the body of the ship came apart. The crew lost consciousness almost instantaneously. They did not even have time to lower their helmet visors.

The survivability group's report included multiple, detailed recommendations on the design of crew-worn equipment, seats and restraints, and the crew compartments for future space vehicles. The report also contained suggestions for crew training and procedures to cope with emergency situations.

NASA's response to the report was positive and supportive. "In the end," Melroy said, "it made me feel better to know that people were saying, 'There are no holds barred—if there's a way of making things better, we'll do it.' It reminded me of what a great culture NASA has—that we are committed to learn everything we can from tragedies."

—

United Space Alliance's Amy Mangiacapra was *Columbia*'s first caretaker and curator, a role she held for ten years. "I maintained the temperature and humidity in the room, made sure everything stayed clean and dust free, gave tours to senators, even scrubbed the floors," she said. "The room is really remote. There's nobody around. I was totally by myself, for eight to twelve hours a day, surrounded by *Columbia*. She was like one of my children. You can't pour that many years into something that special and come away without being completely changed. It was an honor to say that every day I got to go to work and take care of *Columbia*."

NASA transitioned the preservation office curator role from a contractor to a permanent NASA staff position. Mike Ciannilli assumed the lead role, and by January 2016, his charter expanded to include NASA's new Apollo, *Challenger,* and *Columbia* lessons learned program.

KSC strongly urges all new employees—interns, contractors, and civil servants alike—to tour the *Columbia* room with Ciannilli soon after they begin working at KSC. His orientation emphasizes how culture and complacency led to the spacecraft accidents and how individual actions might have made a difference.

Selected artifacts from *Columbia* toured all of the NASA centers in 2008 to reinforce the agency's safety culture. "It's not to point fingers or lay blame," Ciannilli said, "but people need to understand why it's important for every single person to be vigilant every day. You can tell them that, and sometimes it doesn't sink in. But after people see the debris, they say, 'Wow—I really get it now.'"

Being in the *Columbia* room changes people, forcing them to think and reflect. After touring the facility, NASA Administrator and former astronaut—and *Columbia* pilot—Charlie Bolden was visibly shaken. He asked Ciannilli, "How can you do this every day?"

Ciannilli responded, "I try to focus on the good that *Columbia* does in changing people's hearts and minds. I don't do it—*Columbia* does it. She has the loudest voice. She speaks more eloquently than anyone ever could."

The *Columbia* room's 6,800–square foot area does not afford enough space to exhibit every piece of debris. Amy Mangiacapra, Steve Altemus, and Jim Comer selected representative items that are readily recognizable and that tell a story about the accident or the investigation. Educational posters and placards around the room provide details about the items and the recovery and reconstruction efforts. Signed banners—along with hundreds of cards, letters, and posters from schoolchildren—decorate the walls in the room.

An unmarked, locked room off to one side contains the reconstructed crew module. This room is open only to crew families, astronauts, and investigators.

The cockpit window frames, which used to be the first items visitors saw upon entering the room, were relocated in 2016 to the

*Forever Remembered* exhibit at the *Atlantis* building at the KSC Visitor Complex. Now, one first sees the remains of *Columbia*'s air lock and tunnel to the Spacehab module, assembled horizontally on three frames. A rack nearby holds the remnants of maneuvering thrusters.

*Columbia*'s nose landing gear strut lies near the front of the room. A pan underneath the piston still catches occasional drops of hydraulic fluid leaking from the strut. The left main landing gear strut and its tires are also on display, showing how heat entering the left wing affected the inboard and outboard tires differently.

Racks on the right-hand wall hold avionics boxes from *Columbia*'s instrument bay. Some are badly melted and almost unidentifiable. Others are relatively pristine, with serial numbers still clearly legible.

A section of *Columbia*'s vertical stabilizer is surrounded by sections of the right and left pods flanking it on the ship. The pieces clearly tell the tale of how the accident originated on the left wing of the ship. The left maneuvering system pod and left side of the tail are pitted and slumped, with globules of metal melted into the tile. The right side pod has none of that kind of damage.

Lexan frames holding about half of the leading edge of *Columbia*'s left wing exhibit the damage in the vicinity of panels 8 and 9, where the wing breach occurred. Across the aisle is a section of the left wing tile table, containing the recovered tiles from the left wing behind the breach. The tiny "Littlefield Tile"—the farthest west piece of debris recovered—rests on a shelf nearby.

*Columbia*'s nose cap sits on a pallet, still bearing sticks, grass, and dirt from the forest floor from where it was recovered in Sabine County. The engine powerheads that were pulled from mud-filled craters in Louisiana lie on pallets with other pieces of wreckage from *Columbia*'s aft structures.

Starting about one hundred feet back into the room, scores of triple wall storage boxes—bearing labels such as BODY FLAP BOX 1 or PLUMBING-UNIDENTIFIED—line the walls and contain the bulk of *Columbia*'s debris.

Two rooms at the far end of the facility hold the dozens of cryogenic and propellant tanks that were recovered from the field. Some look as though they merely floated to Earth. Others are dented or covered with shredded Kevlar. The lighter objects slowed down rapidly in the upper atmosphere and fell out of the sky. The heavier ones traveled farther and impacted the ground at higher speeds, some still supersonic.

I can't help but be overwhelmed by evidence of the tremendous forces to which *Columbia* was subjected—heating, melting, tearing, shredding, ionization, and impact.

But I'm also amazed at the massive undertaking to bring *Columbia* home and understand how she perished. The display frames for the RCC panels hold large as well as minuscule pieces of the shuttle's wing. Just one frame provides evidence both of the high quality and diligence of the search efforts—that searchers could find pieces of material smaller than a thumbnail out in the wilderness—and the care and skill of our reconstruction engineers, who took these tiny pieces and painstakingly rebuilt the wing. It is akin to assembling a dinosaur skeleton from shattered bits of fossil.

Whenever I visit, I look at the room as a whole and contemplate what it represents. Here lies 40 percent of America's first space shuttle—the collective effort of thousands who designed, built, and maintained her. Here is the vessel that flew 127 women and men on twenty-eight missions into space. Here is the wounded vehicle that fought valiantly to the bitter end to try to bring her last seven crew members home. Here, in this volume the size of two or three average houses, rests the results of the collective efforts of twenty-five thousand people who searched every square foot of a debris field the size of Delaware and Rhode Island combined during three months of 2003. Preserved here is the work of the hundreds of people who processed, cleaned, examined, and cataloged every one of the eighty-four thousand pieces recovered.

Here is a warning of the dangers of complacency and suppressed debate.

And here lies hope people will learn from *Columbia* to make spaceflight safer—although it will never be routine.

## Chapter 14

# THE BEGINNING OF THE END

The dangers inherent in human spaceflight resurfaced only a few months after the *Columbia* accident—while the search-and-recovery operations were still underway. The next space shuttle flight after *Columbia* was supposed to bring home the ISS *Expedition 6* crew of Kenneth Bowersox, Nikolai Budarin, and Donald Pettit in early March of 2003. With shuttle flights suspended indefinitely, NASA and Russia decided to send the *Expedition 6* crew home in May aboard the Soyuz TMA-1 spacecraft docked at the station.

The Soyuz undocked on May 3. A computer error caused the guidance system to malfunction during reentry. The ship went into a ballistic trajectory instead of a controlled descent, subjecting the crew to more than eight times the force of gravity. Ground stations lost communications with the ship when an antenna tore loose during reentry.

The Soyuz descent module landed in a remote area 276 miles short of the targeted landing site.[1] Without a working radio, the crew had no way to contact the recovery forces and let them know where they were and that they were okay. Several tense hours passed before the recovery teams located the ship and extracted the crew. The men sustained minor injuries during their harrowing fall to Earth, but were otherwise unharmed.

The *Columbia* Accident Investigation Board released its report on August 26, 2003. The investigators blamed the loss of *Columbia* as much on NASA's politics and culture at the time as on hardware failure. The report chided the White House and Congress for squeezing NASA's budgets so tightly that safety was at risk. The report cited issues in NASA's transparency, diligence, and oversight dating back to the *Challenger* disaster (and even the Apollo 1 accident in 1967), but which were never fully and permanently corrected in NASA's culture.

The CAIB documented how NASA had permitted the "normalization of deviance" to put both *Challenger*'s and *Columbia*'s crews in harm's way. The *Challenger* accident was the result of a known systems issue in the shuttle's solid rocket boosters that had not been corrected. There had been partial burn-through of the O-rings on several previous missions—including the second flight of the space shuttle—but the details of the problem and the potential catastrophic outcomes never came to the attention of the launch decision makers. Lower-level engineers and managers did not allow the issue to be brought forward. Their desire to meet the launch manifest—and each organization not wanting to be the reason for standing down—trumped sound engineering practices of full and open discussion. Human and organizational failures doomed *Challenger* just as surely as the O-ring failure.

Similar conditions led to the *Columbia* accident. Foam shedding from the external tank was not within the design specifications for the space shuttle, but it had happened repeatedly over the years. Based on the shuttle's demonstrated ability to survive hits from launch debris, managers justified continuing to fly while pursuing a new design. A mission four months before *Columbia*'s flight also suffered damage from external tank foam, and yet the issue was not even addressed at *Columbia*'s flight readiness review.

Clearly, changing organizational culture—and making those changes stick—is much harder than improving technology.

We were working on fixing the foam problem, but in hindsight, not nearly as aggressively as we should have been. NASA chose to press on in order to meet the unrealistic and self-imposed deadline of completing the core of the ISS by February 2004. The urgency to finish the ISS overrode the urgency to fix a potential safety issue.

NASA's decisions—and nondecisions—ultimately caused the loss of *Columbia*, took the lives of her crew and two searchers, endangered citizens on the ground, resulted in the expenditure of hundreds of millions of dollars for a recovery and reconstruction effort, and delayed ISS assembly missions for three years.

Senior NASA officials expressed surprise throughout the investigation as they learned about the concerns people said they had tried to raise while *Columbia* was in orbit. Leaders said they had no idea serious issues were not being elevated to their attention, when policies were clearly in place to encourage open and honest discussion. As with *Challenger*, the agency's culture eroded over time into one of "prove to me why it's *not* safe to fly." It created a fear to speak up and be a dissenting voice, which ultimately stifled debate and killed the crew. Administrator O'Keefe also said that he was disappointed to learn that no one had called a safety hotline or alerted high-ranking officials about their concerns—a system that was already in place to allow anyone to escalate issues anonymously and without fear of retribution.[2]

In addition to the deeply embedded cultural issues at the agency that still needed to be fixed, the CAIB pointed out that the shuttle's design was inherently flawed. Too many problem scenarios were possible from which a shuttle crew had no way to escape or survive. The risk could be mitigated somewhat, but in the CAIB's opinion the Shuttle Program was "operating too close to too many margins."[3] The CAIB recommended that NASA accelerate steps to replace the space shuttle.

Administrator O'Keefe embraced all of the CAIB's recommendations and assured the board NASA would implement them. It was tough medicine to take, but we needed it.

All told, the *Columbia* recovery, reconstruction, and investigation cost two lives and $454 million. Of that, FEMA spent $302 million for public safety and the search operations in Texas and Louisiana. NASA's $152 million share of the cost included the recovery and reconstruction of the debris and the funds needed to support the CAIB's investigation.[4]

—

As a sidebar to the investigation, the CAIB quietly asked us to determine if there might have been an opportunity to launch a rescue mission to save *Columbia*'s crew, had we known that the ship was doomed. At the same time, they asked engineers in Houston if there might have been a way for *Columbia*'s crew to repair her wing with materials on board the ship. Admiral Gehman deliberately waited until May to request the analyses to allow emotions to cool down following the accident. We had already conducted our own internal studies before Gehman's request.

When *Columbia* launched on January 16, her sister ship *Atlantis* was in the Orbiter Processing Facility hangar. She was almost ready to be mated to the external fuel tank and solid rocket boosters that were already stacked in the VAB. Could we have gotten *Atlantis* off the ground in time to save *Columbia*'s crew?

After the accident, I studied how we could have accelerated processing activities and eliminated tests without jeopardizing the safety of *Atlantis* and her crew. For example, we could skip the terminal countdown demonstration test and the cryogenic fuel loading tests, shaving several days off the schedule.

My analysis showed that the rescue scenario was feasible from the KSC processing and launch perspective—but only if we got the "Go" by January 23. For that decision to be successful, we would have already needed to be in high gear immediately after learning about the foam impact on *Columbia*'s wing—significantly altering the crew's on-orbit activities starting on January 20. NASA would have needed detailed images of the wing from America's intelligence assets, or would have had to send some of *Columbia*'s crew outside to inspect the wing. That space walk would have needed to happen on the second or third day of the mission—a completely unrealistic assumption given mission timelines and goals. For all intents and purposes, the mission would have been over at that point, whether or not the wing was actually damaged.

With the ship confirmed to be mortally wounded, mounting a rescue mission would have been a mammoth undertaking and a very risky proposition. Knowing that foam from the external tank had doomed *Columbia*, would we dare to launch *Atlantis* with an identical external tank, possibly one with the same fatal flaw? If *Atlantis* were also damaged during ascent, we would have lost two shuttles and two crews.

This would not have been solely a NASA decision. President Bush would have needed to be involved in the process, and there would have been very little time for debate.

*Columbia*'s crew would have been told to shut down all noncritical activity on the ship and sleep extended hours to prolong the cabin's carbon dioxide removal capability while awaiting the rescue. They would also close down the Spacehab module, effectively ending *Columbia*'s science mission.

Meanwhile, the rescue mission crew would rehearse procedures in Houston for rendezvousing with *Columbia* and transferring her crew to *Atlantis*. At KSC, a "full court press" of round-the-clock activity would put *Atlantis* on the launchpad no earlier than January 31.

The earliest possible launch date was February 11, assuming that *Atlantis* would not be equipped with its remote manipulator arm. If the arm was going to be installed—and it was almost certainly needed for the rescue mission—the earliest launch date slipped to February 13.

If all went well, *Atlantis* would have rendezvoused with *Columbia* and kept station with her, with the ships' open payload bays facing each other. The rescue crew of four would run a tether between the two ships, and then bring *Columbia*'s crew over to *Atlantis*, one by one.

Once the transfer was complete, *Atlantis* would head home with eleven people. Four of them would have to be strapped to the deck in the crew module during reentry, since the shuttle only carried seven seats. NASA would command *Columbia* to reenter the atmosphere, timing the maneuver to have the ship burn up over a remote area of the South Pacific.

If everything went according to plan, *Columbia*'s crew would have had about a two-day margin in their consumables. They would have been in orbit almost a month by that point.

Everything hinged on making the momentous decision on January 23, following the decision to conserve the air scrubbers on the fourth day of the mission.

And remember that the request for intelligence imagery surfaced on January 22. Even if the request had been approved at that point, we wouldn't have had the pictures in time to make an informed decision. Furthermore, the reduction of the crew's normal activities to conserve consumables would have made a space walk unfeasible in the first place.

It had been already too late for a rescue.

—

Repairing *Columbia* in orbit was an even more uncertain proposition. The crew lacked suitable repair materials or equipment. *Columbia* did not have its remote manipulator arm installed, which would have been needed to support astronauts during a space walk. Even if by some miracle the crew could pack the hole in the wing with ice and fashion a metal cover for it—two options that engineers explored—it appeared highly unlikely that this would sufficiently protect the ship during reentry.

When we saw the analyses, there was no grumbling, but there was grief. We couldn't save the ship. *Columbia* was doomed, no matter what. Maybe we could have saved the crew. But there were so many what-ifs and assumptions, so many things that had to go completely differently from the very first hours of the mission. Would it have been successful? I don't know. We never even had the chance to try.

As much as it hurt people to think about the remote possibility of saving *Columbia*'s crew, the study helped prompt discussions on how to save a future crew of a damaged shuttle.

Missions to the ISS had the advantage of delivering the crew to a place where they could wait for a subsequent mission to retrieve them or go home via the Soyuz. Assuming an injured shuttle could dock to the ISS, its crew could await a rescue mission for ninety days or more.

This "safe haven" capability was one of the key factors that led NASA to approve the resumption of shuttle flights.[5]

The issue was more problematic for servicing missions to the Hubble Space Telescope, which is in a different orbit than the ISS.[6] Because of the laws of orbital mechanics, the amount of fuel needed to move the space shuttle between the orbits of Hubble and the ISS was far greater than the orbiter could carry. The ISS could not be a safe haven for a Hubble mission. Without a rescue capability, Sean O'Keefe felt that the risks to human life did not justify prolonging Hubble's life by a couple of years. On January 16, 2004, he canceled the final planned Hubble servicing mission.[7]

Mike Griffin replaced O'Keefe as NASA administrator in April 2005. Griffin believed it was so important to extend Hubble's life and capabilities that he was willing to reinstate the servicing mission—provided the external tank foam shedding issues were resolved and adequate crew rescue capability existed. KSC and JSC used the *Columbia* rescue scenario to design a one-time rescue mission that could back up the Hubble servicing mission. Griffin formally approved the servicing mission after the successful completion of STS-121 in July 2006.[8]

On May 11, 2009, *Atlantis* was poised for launch to the Hubble from Pad 39A at Kennedy. Standing on Pad 39B two miles to the north was *Endeavour*, ready to go into orbit if any problems occurred with *Atlantis*. For the first and only time, NASA had two shuttles in launch countdown simultaneously. We were ready to launch *Endeavour* one day after *Atlantis* if necessary. Tremendous dedication and work went into getting us to this dual-launch posture. Fortunately—like many other things in the space business—this contingency capability was assured but never needed.

*Atlantis*'s flight went flawlessly, so the rescue mission never flew. *Atlantis*'s crew successfully prolonged Hubble's life and upgraded its instrument package.

In a roundabout way, what we learned from the *Columbia* accident had once again contributed to the advancement of scientific discovery.

—

*Columbia*'s STS-107 accident did not end the Space Shuttle Program. However, it informed the decisions that did. As brilliant as the space shuttle's technology was, the vehicle could never be made acceptably safe to risk further flights after the International Space Station was completed.

Even before the CAIB issued its report, debates about shuttle safety raged within NASA. "That is the pattern in any aftermath of a cathartic event like this," Sean O'Keefe said. "There will be plenty of things that are emerging now and will continue to emerge that will motivate a change in the way we look at doing business."[9]

Fundamental questions resurfaced. Did the benefits of sending humans into space outweigh the risk and expense? Maxime Faget—the legendary engineer who designed America's Mercury space capsule and managed the design of every other American manned spacecraft—felt that the country should immediately halt all human spaceflights until a safer vehicle could be built.[10]

Had it not been for America's commitments to its international partners to complete the ISS, the Space Shuttle Program could very well have ended with the loss of *Columbia*. Building the ISS was a matter of international treaty.[11] We had to see through an endeavor into which the world community had invested tremendous time and resources. Without the shuttle, there was no other way to get the ISS modules into orbit and assemble the station.

Some within NASA naturally felt hesitant to take further risks after the accident. Dave King said, "It shakes your confidence. It shakes every part of you, when you're part of making decisions that kill your friends and make your friends suffer." Overcoming the grief and the mood of shared guilt for having let the crew down proved a tough management challenge. Leaders concentrated on rebuilding people's faith and pride in their work. Ultimately, a driving motivation was the perceived obligation to *Columbia*'s crew to carry their mission forward—to ensure that their sacrifice had not been in vain.

In May 2003, NASA named former astronaut Thomas Stafford and retired shuttle astronaut Richard Covey to head an independent task force to evaluate our plans for returning the shuttle to flight. The Stafford-Covey Task Force would provide an ongoing, unbiased assessment of how NASA was implementing the CAIB's recommendations. As the time for the next mission approached, the task force would advise the NASA administrator whether they felt everything had been done to make the shuttle safe to fly again.[12]

While the space shuttle was being recertified for flight, NASA moved forward with a new "Vision for Space Exploration," announced by President George W. Bush on January 14, 2004. The Vision called for completing the ISS and retiring the space shuttle by 2010. Meanwhile, NASA would develop the new Constellation Program, which included expendable launch vehicles and the Crew Exploration Vehicle (Orion), which was a capsule like Apollo, but on a larger scale. Crews would begin flying on Constellation in 2014. Constellation would return Americans to the Moon by 2020 and put them on Mars in the not-too-distant future.

Announcing the end of the Space Shuttle Program felt like a crushing blow to many in the NASA community. We had planned to fly the orbiters until at least 2020. Now only a handful more than twenty missions would be left before we retired the shuttle. United Space Alliance's Larry Ostarly said, "Quite frankly, I don't know what made me sadder—that we lost *Columbia*, or that it cost the program."

—

In late 2003, program managers and leaders at the NASA centers began identifying the changes to hardware, processes, and practices to address the findings in the CAIB report, as well as other findings that came to light during the investigation. Engineers at Marshall Space Flight Center reeducated the workforce in the proper procedures for applying foam to the external tank. Johnson Space Center engineers developed hardware and techniques for inspecting and repairing a shuttle in orbit. The first several days of a mission would now include a complete inspection of the shuttle's heatshield, both by the crew on the shuttle and by astronauts inside the International Space Station, who would photograph the shuttle as it approached. We overhauled our inspection and processing procedures at Kennedy and significantly improved the tracking camera coverage for future launches. Redesigned flight plans allowed launching only in daylight hours, to allow cameras to monitor the ascent phase of the mission. Critical phases of reentry could fly only over unpopulated areas. Mission Management Teams would be required to meet every day the shuttle was in orbit, no matter how well things seemed to be going. Roy Bridges established NASA's Engineering and Safety Center at Langley Research Center, providing a resource any engineer could call with a concern that they believed needed to be examined.

Shuttle Program managers implemented sweeping changes to address the organizational culture issues the CAIB identified. Mission managers would be required to attend specialized training sessions about how to foster full and open debate on any issue. Searching out dissenting opinions became the norm and was embraced throughout the program. No longer would lower-level employees feel reluctant to speak up if they had an issue or alternative opinion.

I can recall post-*Columbia* program-level meetings where we could not adjourn until at least one dissenting opinion was presented. It was a little awkward, but it was the right thing to do—to really show the team we meant what we said about open discussions.

Implementing the space shuttle safety recommendations took two years. Meanwhile, the International Space Station remained manned, albeit with small crews who launched to and returned from the ISS aboard Russian Soyuz spacecraft.

Eileen Collins and her crew trained for the "return to flight" STS-114 mission of *Discovery*, which would take a logistics module loaded with cargo to the station. The crew's primary mission, however, was to demonstrate improvements in shuttle safety.

Collins had intended that her mission—originally scheduled for March 2003—would be her final one. She planned to retire from the astronaut corps that summer and move to Florida with her husband, who was a pilot for Delta Airlines. When the *Columbia* accident occurred, she knew that her plans would have to change. "It would just look bad if the commander of the next mission retired, no matter what the reason. People would think I had lost faith or was worried about my safety," she said. "I was not going to do that to NASA, no matter what my personal plans were, because I had confidence we could fly our mission just fine."

NASA invited Roger and Belinda Gay, Marsha Cooper, Terry Lane, and many of the people from East Texas who had been so helpful in the *Columbia* recovery effort to come to KSC and witness *Discovery*'s launch. It would be a fitting tribute for these people to see the shuttle fly again as a result of their hard work and sacrifices. The group toured Kennedy's facilities, but their hope to see the shuttle lift off were dashed when a fuel tank sensor problem scrubbed the planned July 13, 2005, launch. They had to return home before our next launch attempt.

We could not locate the cause of the sensor problem. After a week, engineers determined it was not critical to flight safety, so we set a new launch date of July 26.

As the countdown came out of the T minus nine-minute hold, I cleared *Discovery* for launch and told the crew, "On behalf of the many millions of people who believe so deeply in what we do—good luck, Godspeed, and have a little fun up there!"

*Discovery* finally lifted off the launchpad—907 days after the *Columbia* accident.

Trouble ensued almost immediately.

A large bird struck the fuel tank less than three seconds after liftoff—which fortunately caused no damage to the vehicle. A small piece of tile fell off the edge of the shuttle's nose landing gear door some time before the solid rocket boosters separated. One edge of a thermal blanket under the commander's cockpit window also came loose. And to everyone's horror, the external tank shed several large pieces of foam, one of which was about half the size of the piece that fatally wounded *Columbia*. The largest piece fortunately missed *Discovery*, but another piece of foam struck the shuttle's right wing.

On-orbit inspection of *Discovery*'s heatshield revealed only minimal damage. Wind tunnel tests showed that the loose insulation blanket would not cause a problem. However, the extent of foam shedding from the external tank was absolutely unacceptable, since the issue had supposedly been fixed.

While *Discovery* was still in space, NASA declared a moratorium on future shuttle flights until the foam shedding problem was resolved.

The mission itself went smoothly. *Discovery* docked with the ISS and delivered much-needed cargo and supplies. In three space walks, the crew demonstrated shuttle tile repair techniques, replaced a failed gyroscope on the ISS, and installed an external stowage platform for ISS tools and equipment. Then astronaut Stephen Robinson conducted an actual repair on the shuttle, removing two gap fillers that were protruding from between tiles on *Discovery*'s belly. It was the first time an astronaut had ever ventured underneath the shuttle during a space walk. "Other than the launch, that was the riskiest thing we did on this mission, because we hadn't trained for it," Collins said.

*Columbia*'s crew remained in the hearts and minds of the return-to-flight mission. The STS-114 crew had redesigned their crew patch to incorporate the STS-107 mission emblem. Collins kept a photo of the *Columbia* crew on display in *Discovery*'s flight deck throughout the mission. "Whenever we were up on the flight deck, we had that crew with us," she said.

On August 3, the shuttle and ISS crews gathered to send birthday greetings to Matthew Husband, son of *Columbia*'s late commander. The crews also read a short memorial service composed by astronaut Andy Thomas, entitled "Exploration—The fire of the human spirit, a tribute to fallen astronauts and cosmonauts."[13] Collins concluded the service saying, "For all our lost colleagues, we leave you with this prayer, often spoken for those who have sacrificed themselves for all of us:

They shall not grow old, as we that are left grow old:
Age shall not weary them, nor the years condemn.
At the going down of the sun and in the morning
We will remember them."

*Discovery* was supposed to land at Kennedy Space Center on August 8. Bad weather at KSC forced NASA to wave off two landing opportunities that day and two the next day. NASA finally directed *Discovery* to land at Edwards Air Force Base in California.

Collins and her crew felt absolutely confident about the final phase of flight after the extensive inspections they conducted during the mission. "As a habit, I called out Mach numbers so that the astronauts in the mid-deck knew where we were in the reentry profile," she said. "None of us said anything about it, but we were all aware when we passed the airspeed and altitude where *Columbia* had her accident." Collins brought *Discovery* in for a landing and called "Wheels stop" at 8:12 Eastern Time on the morning of August 9, 2005.

A month later, Hurricane Katrina slammed into New Orleans.[14] Shuttle external tank assembly operations at NASA's nearby Michoud Assembly Facility were suspended for nearly two months while NASA repaired wind and water damage to the facility. In November, NASA inspected one of the external tanks that had twice been filled with and drained of liquid oxygen and liquid hydrogen. New scanning techniques revealed cracks deep in the insulating foam that were not visible on the surface. It appeared the foam cracked as the tank contracted and expanded due to thermal changes. NASA realized that it was an engineering issue—not human error in applying the foam—which had caused the foam shedding problem.[15] Shuttle Program Manager Wayne Hale said shortly thereafter, "I flew to New Orleans within a few days, and called an all-hands meeting where I publicly apologized to the foam technicians. They had not caused the loss of *Columbia* through poor workmanship. Those guys were reeling from the hurricane's devastation to their homes and community, and had lived with nearly three years of blame."[16] NASA could now make the needed design changes to the tank.

STS-121—the second post-*Columbia* return-to-flight mission—launched on July 4, 2006. The external tank only lost a minor amount of foam, and it occurred after the most critical time during ascent to orbit. The flaw that had doomed *Columbia* was finally fixed. Much to many people's amusement, bird droppings seen on the shuttle's right wing several days before launch were detected in the on-orbit inspections.[17] Otherwise, the mission was nearly flawless.

The shuttle had come roaring back.

—

Between 2006 and 2011, the three remaining shuttles in America's fleet flew twenty more missions after STS-121. Their crews completed the International Space Station and serviced the Hubble Space Telescope one final time. There were no other accidents or close calls during the rest of the Shuttle Program. NASA's diligence following the *Columbia* accident paid off.

The Constellation Program had been making slow and steady progress, but its only flights were a test of the Ares I-X rocket in 2009 and a test of the launch escape system in 2010. As has happened all too often with NASA's budgets over the years, the agency did not receive the funding it needed to realize its ambitious vision. In 2009, a presidential commission reported Constellation to be so far behind schedule, over budget, and underfunded that it was impossible for the program to meet any of its goals. The administration removed Constellation from NASA's fiscal year 2010 budget, effectively canceling the program. Meanwhile, NASA had already been moving forward with the termination of the Space Shuttle Program after STS-134.

With Constellation canceled and the shuttle winding down, NASA was in a bind. The Commercial Crew program was born, calling for private companies to build vehicles and operate flights to the ISS under NASA charter. However, the program was still in its infancy, and the first commercial crew flights were at least four years away. NASA decided to extend the Space Shuttle Program with one final mission to carry supplies, equipment, spare parts, and the Alpha Magnetic Spectrometer to the ISS. But the transfer of decades of knowledge and talent to the subsequent program—a critical success factor in the early days of America's manned spaceflight programs—would not be possible.

*Atlantis* rolled out to Pad 39A on June 1, 2011, for STS-135. Many people at KSC had spent their entire careers working with the space shuttles. It seemed impossible to believe this would be the final mission.

I ruffled some feathers by openly addressing the issue with the Firing Room personnel—most of whom were contractors—at the conclusion of the launch simulation. Most members of our launch team had been together for twenty or more years. Many of these wonderful, dedicated people had taken part in the search for *Columbia* or had worked in the reconstruction hangar. Virtually all of the contractors were going to be laid off when the STS-135 mission ended, and no one had addressed the eight-hundred-pound gorilla in the room. So I told my team what I really thought. As politically inappropriate as it was, I apologized to them on behalf of NASA. I got spanked for that, but I felt it was the right thing to do for my team.

We launched STS-135 on July 8, 2011. *Atlantis* returned to KSC in a predawn landing twelve days later. Completion of the ISS was cause for celebration, but it also meant the end of the Space Shuttle Program. LeRoy Cain and I hugged each other tearfully on the Shuttle Landing Facility runway. We had both been through so much since the terrible events of February 1, 2003.

With no new system to transition into, opportunities for contractors to stay in the space program were scarce. Layoffs began the next day. This had a far-reaching effect on the teams that developed over the thirty-year life of the Space Shuttle Program. Losing this extraordinary expertise was a casualty whose impact cannot be fully appreciated until the time comes to rebuild it. Pam Melroy said, "We accumulate wisdom. The hardware is the least of it. Someone could hack into your computer and steal all your information about the shuttle, but they wouldn't have the slightest clue how to operate it. It's all about the corporate knowledge that we share."

After the mission, I told reporters, "It doesn't matter what the change is—any major change in one's life, you go through these four stages: denial, anger, exploration, and acceptance. We've all been through that now in the Shuttle Program and we've accepted the fact that it's over. This is the end of the program, and people will move on and do well." I concluded my comments with, "It's important, but it's not the end of the world. The sun will rise again tomorrow."[18]

As a career civil servant, I was assured of a job after the Shuttle Program ended. For several years, though, I had been toying with the idea of retiring. With no manned launches in the foreseeable future, I thought this was as good a time as any to punch out. The launch director role fit me pretty well. After a career of launching shuttles, I don't think I could have been a budget guy behind a desk. I retired from NASA in November 2011.

NASA's remaining shuttles retired to museums in 2012—*Atlantis* at Kennedy Space Center's Visitor Complex, *Discovery* at the Smithsonian's Udvar-Hazy Center in Virginia, *Endeavour* at the California Science Center in Los Angeles, and *Enterprise* at the Intrepid Sea, Air, and Space Museum in New York City.

Chapter 15

# CELEBRATING 25,000 HEROES

As time passed following the accident and the remarkable events of the first half of 2003, people gradually returned to their daily lives. No one wanted to dwell on the horrors of the accident or the tremendous hardships that people endured in the immediate aftermath. And yet, there was a collective need to honor the sacrifices the crew made on behalf of their countries and to celebrate the incredible work accomplished by thousands of people in the aftermath.

Just a few months after the recovery effort wrapped up, the Texas communities along *Columbia*'s path began thinking about how to honor the crew of *Columbia* and tell the story of their communities' roles in the recovery.

Lufkin placed the commemorative plaque presented by NASA at the farewell dinner in the city's Louis Bronaugh Park, installed a monument, and flew a *Columbia* flag in the park's flag circle. Displays in the town's Civic Center highlighted Lufkin's key role in the recovery effort. Farther east, San Augustine placed a stone memorial to *Columbia* at the town's Civic and Tourism Center.[1]

Hemphill in particular sought to memorialize both its identity as "ground zero" for the accident and the actions of the thousands of volunteers who helped NASA search for the crew. In a very real sense, Sabine County's citizens believe their community is hallowed ground—the place where God chose to bring *Columbia*'s crew to rest.

Belinda Gay, Marsha Cooper, and Ellen Mills took the lead and devoted years to envisioning and promoting several commemorative sites for *Columbia*. First, the town revamped the raised circular Lone Star monument at the intersection of Highway 87 North and Farm Road 83. Its new design incorporated the STS-107 mission emblem in the center of the star and placed the motto THEIR MISSION BECAME OUR MISSION around the outer ring of the circle. The site also included stone monuments to the *Columbia* crew and the two searchers who perished in the helicopter accident, as well as the US and Texas flags and the *Columbia* banner.

Second, they explored developing a commemorative glade at the "nose cone site" west of town. The community solicited site design concepts from architecture students at Texas A&M University. Hemphill's memorial committee then spent several years working to persuade legislators to designate the site as a national landmark or a national park. Their proposal made it to Washington, DC. However, the National Park Service determined in 2014 that despite the historical significance of the *Columbia* accident, the site did not meet the suitability standards to become part of the national park system.[2]

For the third planned memorial, Gay, Cooper, Mills, and other citizens of Sabine County envisioned a museum in the town that would serve as a visitor center, a repository for artifacts and information related to the STS-107 crew and the community's role in the recovery, and an educational center that would eventually include a space shuttle cockpit simulator. Generous donations from many people around the country—including a major grant from Mr. Albert Smith in memory of his wife, Patricia Huffman Smith—funded the museum.

NASA provided educational materials. Lockheed donated unused shuttle wing RCC panels and a shuttle nose cap. The families of *Columbia*'s crew shared personal mementoes from their loved ones. Evelyn Husband donated Rick's contact lens case—the one that Marsha Ivins and Jim Comer had examined in the reconstruction hangar.

Laurel Clark's husband Jon donated her collection of science and aviation books, filling the shelves in the museum's education center.

The night before the museum's dedication, Willie McCool's father approached Marsha Cooper and asked if she could open the display case containing his son's memorabilia. The senior McCool removed from his pocket a pair of gold astronaut wings—which would have been presented to *Columbia*'s pilot upon completion of his first space mission—and placed them in the case. "This is where they belong," he told her.

The Museum opened on the eighth anniversary of the accident—February 1, 2011. Sabine County's commemoration committee produced a video, *Of Good Courage*, on the tenth anniversary of the accident. Both the museum and the video provide heartfelt evidence of the determination of the people of Sabine County to ensure that *Columbia*, her crew, and the remarkable accomplishments of the citizens of East Texas will not be forgotten.

In NASA's Apollo, *Challenger*, and *Columbia* Lessons Learned Program, Mike Ciannilli sought opportunities to educate the broader public on the sacrifices made by astronauts and their families on behalf of the country's space program. As part of his educational mandate, he worked quietly behind the scenes with the families of the *Challenger* and *Columbia* crews to obtain their consent to present representative debris from the two vehicles in an educational display at KSC, along with personal memorabilia from the fallen astronauts. Thanks to his efforts, part of *Challenger*'s fuselage sidewall and *Columbia*'s cockpit window frames were enshrined in 2015 in the "Forever Remembered" exhibit in the *Atlantis* building at the KSC Visitor Complex. They are the only artifacts from either vehicle viewable by the general public.[3]

Pat Adkins now volunteers as a docent in the *Atlantis* building. When the exhibit opened, Adkins was somewhat startled to see the frames on display, still bearing embedded bits of the dried mud and grass from when he retrieved one of them in Sabine County.

In quiet times, he finds himself at the exhibit, reflecting on his experiences in East Texas after the accident. "It's not a tourist attraction. It's a memorial. People need to understand that there is a cost of doing what we did and what we're going to continue to do. The public needs to see that there are people out there doing these things—not only for themselves, but also for their country—and it's a good thing. This is a reminder that sometimes it's hard."

—

What did the participants throughout this mammoth undertaking learn about themselves, their communities, and their country as a result of their role in bringing *Columbia* home? And what should Americans in general take from this moment in our history?

Astronaut Jerry Ross said, "First and foremost, people need to understand the greatness of the United States of America and its citizens. The outpouring of support and prayers we received was tremendous. Second, the United States has an incredible wealth of capabilities. To see the energy and expertise and materials and technical capabilities that descended on Lufkin within hours of the accident was so reassuring. Finally, there were no ulterior motives. Every individual was there to do what they could to get the country's space program flying again."

Local communities proved the usefulness of the incident command system. I suspect that most Americans are not aware their community leaders have been trained in this powerful process for dealing with disasters, both natural and man-made. In Texas, every county judge is required to attend a three-day class on the system, as are most firefighters and law enforcement personnel.

Mark Allen was the logistics lead on the ICS team headquartered in Hemphill during the recovery. He is now county judge of Jasper County. He said, "The training's important, but it's more than just checking off that box. It's the network you build through exercising and training and working together. All of us from the local counties have banded together on different incidents, and we've learned to trust each other's judgment. That's real important in a critical situation where it's high stress, no sleep, and lots of coffee, and you're making it up as you go along. You've got a structure and a whole bunch of other people that you know that can do this."

The disaster provided the local communities an opportunity to be their best—to demonstrate how their leaders and citizens could step up when their country needed them most. "If people couldn't do one thing, they'd find something else to do that would help out. Everybody did it as a team, and that was the reason for success," said Greg Cohrs.

FEMA's Scott Wells said, "The shuttle disaster is a great case study on how the whole community works together to accomplish something. The president's 2010 national security strategy was that the Government can't do it all alone anymore. The threats that face the nation are so big that it's going to take the whole community—local, individuals, volunteer organizations, states, Federal government—everybody. That is starting to take traction now."

As Jerry Ross said, another key to success was that everyone operated without personal agendas. More than 130 agencies and more than 300 volunteer groups and private organizations worked together on the recovery. While we may think of agencies as faceless entities, they are in fact composed of everyday people who are trying to do their best. Rather than carving out their own professional turf, personnel in the *Columbia* aftermath concentrated on how they could bring their agencies' resources and expertise to bear to help solve problems and get things done.

"I don't care how much you fuss and fight—you can be brought together on common grounds," said US Forest Service law enforcement officer Doug Hamilton. "But I never thought it would be to this extent. I don't think anyone will ever be able to tell the story of just how it was, because you had to be here to see this—how everybody worked together without arguing and bickering." Hemphill City Manager Don Iles said, "As tragic as this was, it was a good moment for our community. Everybody in this county did the right thing."

In contrast to the *Challenger* accident investigation—which NASA held relatively close to the vest—NASA's leadership after the *Columbia* accident was much more open about rooting out the issues. Sean O'Keefe demanded that we bring everything out into the open as rapidly as possible ". . .warts and all. You're going to hear about it sooner or later. So let's hear it sooner so that we can deal with it. I take great comfort that of the many stories about the accident that were written after the fact, not a single one of them suggested that we were suppressing information."

As a result, people from all walks of life seemed even more eager to help NASA return to flight. KSC's Stephanie Stilson said of her experience with the people of Nacogdoches, "We had screwed up. And to know that these people were still willing to support us and be part of resurrecting our reputation made me appreciate even more being a civil servant. I'm proud of working for NASA and the federal government and being an American citizen."

In its darkest hour, NASA reconnected with the country at a local level, uniting with people who might never otherwise have had contact with the agency. Astronaut Brent Jett said, "It made me so much more appreciative of the unsung, good Americans who will respond and do everything they can to help. Their true goodness was humbling. That was the biggest thing that affected me."

Dave King said, "I learned it wasn't *my* space program. All these people cared deeply about what NASA was doing, our successes and failures and the tragedy we had. But they wanted us to understand that it's *America's* space program—it's not NASA's program or that of the individuals who work for us or our contractors."

Astronaut Dom Gorie added, "This kind of response made it so very clear that this country absolutely demands a manned space program. They want to continue it and honor the people that had given their lives to it. When it came time for me to think about my next mission—and was I truly willing to put my family through this kind of stress again—the support we got was a big part of my decision to fly again."

Leaders were needed at all levels of the recovery and reconstruction efforts. People stepped up to the task, and in many cases greatly exceeded even their own ideas about what they were capable of doing. Jim Comer was one example. He was a good leader before the reconstruction, but he was *great* by the time he left. He showed us a confidence that you can't learn from classes. And there were so many others. Amy Mangiacapra rose way above her pay grade to do the things she did. Ed Mango, Steve Altemus, and many others saw their careers deservedly accelerated as a result of their leadership in a crisis situation.

Jamie Sowell of the US Forest Service said, "I learned so much from leading and inspiring a group of volunteers—getting 110 percent out of them every day, asking them to do something after the glamour has worn off and they want to go home. 'Come on! Just one more ridge. One more road. One more hour.' Something great was going on here. Everyone wants to be part of something great. Finding the part that they can be successful in is the key—positive ways that they can contribute. I've carried those lessons every day since then."

Jon Cowart, who now works with NASA's Commercial Crew program said, "While I had great appreciation for the technical things we did in the hangar, I learned to value the people on my team much more. I was much better able to relate to the people working under me and with me. It made me a better manager of human beings."

Security special agent Linda Rhode said, "The astronauts gave everything they had. They have a lot of courage, and they have to work together in a very small space. If they don't, their mission won't succeed. When I'm working on projects with difficult people, I try to remember, 'What would the astronauts do?' They would take the high road. What's best for the agency? What's best for our mission? Those are the important questions."

Every person interviewed for this book said that the *Columbia* experience was a singular defining moment in his or her life. Firefighter Jeremy Willoughby, who was on the fire crew that found the OEX recorder, said, "I come from a very small town—Madison, Florida—and it's just unbelievable to think when I look back on it. How did I get put on this huge part of history?"

Mike Ciannilli said, "It was a time in my life I'd never want to experience again. But I can't ever imagine not living through it."

Volunteer searcher Dwight Riley from Sabine County said, "It's the most rewarding thing I've ever done. I'm seventy-eight now. If they called me to do it again, and if I was still able to walk, I'd be right there."

Scott Thurston, now working with the Orion and Space Launch System programs at KSC said, "People ask me how I can stay so calm under pressure. Compared to standing in front of the orbiter you and your friends have worked on for all those years, and seeing it broken apart and trying to investigate what happened, this is a lot easier to handle. I stay calmer in adversity than I used to."

Robert Hanley, who had worked so closely with *Columbia*'s crew, said, "It caused me to look long and hard at my life, my career, and where I was headed. It taught me that when you get knocked down, you have to get up and keep working hard. Bad things happen. You deal with them and keep going. You need to take time and cry, and that's okay. There's nothing wrong with grieving. It needs to happen and you can't do it alone."

Citizens of East Texas had a unique opportunity to work with and get to know individual astronauts, and they were impressed at their commitment and their humanity. Texas Forest Service air coordinator "Boo" Walker said, "We had one meeting where I was probably the only one at the table who wasn't an astronaut. A question came up and everyone looked at me and said, 'What do you recommend?' These people will listen to you. They might not always agree with you, but they truly want to hear your opinion."

Cecil Paul Mott, Hemphill's electrical supervisor, said, "We learned that these people we see on television climbing into rockets are more than just faces. Now you think of the fragility and the humanity of what's out there. These people, like the explorers of our past, endured what they did so that we can walk where we're walking today."

Greg Cohrs wrote, "I feel a special connection to the crew and to NASA. I will carry that until the day I die. I am very thankful that

God allowed us to recover the crew, as I know I would have been tormented until my passing had we not recovered the entire crew."[4]

The FBI's Terry Lane said, "Joshua 1:9—the verse that Rick Husband read to his crew before the mission—became the watchword for the whole recovery. 'Have I not commanded you? Be strong and courageous. Do not be afraid; do not be discouraged, for the LORD your God will be with you wherever you go.' After *Columbia*, I made several trips to Iraq and Afghanistan for the FBI. I had a dog tag made up with that Scripture. I had it with me every time I went over there."

Marsha Cooper of the US Forest Service said, "It made me realize how much amazing love is really there. I never felt so much love for humans or mankind."

—

The *Columbia* accident, with the loss of its seven crew members and the two searchers, was a profound tragedy, but many people felt that divine intervention prevented things from being worse than they were. Had *Columbia* disintegrated two or three minutes earlier, much of its debris would have fallen on Dallas and its suburbs, causing untold damage. A breakup a few seconds later would have sent some of the crew members' remains into Toledo Bend Reservoir or the Gulf of Mexico, from which they would likely never have been recovered. If NASA had waved off the first landing attempt, *Columbia* could have fallen into downtown Houston on its next orbit.

Had the accident occurred in the spring or summer, the East Texas "Pine Curtain" would have been impenetrable because of the new growth of underbrush and briars. Heat, humidity, snakes, and alligators would have taken a toll on the searchers. Had the accident occurred during fire season rather than February, very few crews would have been available to assist with the search.

The citizens of East Texas feel the hand of divine providence in bringing *Columbia* and her crew to rest in their community. It afforded them the opportunity to "love thy neighbor" by comforting the NASA family in their time of grief.

One trusts that the residents of any region in America would have responded with the same grace, love, and dedication that the communities of Texas displayed.

No exact accounting was ever made of the number of Americans that helped NASA find *Columbia* and her crew, return them home, and reconstruct the orbiter from the debris. That would not have been possible, given the nature of the operation, the large number of agencies involved, and the number of volunteers who showed up to help for a few days or hours and then moved on. The consensus is that about twenty-five thousand Americans were involved in one way or another. Some stayed on through the entire duration of the effort, some for briefer periods. But all helped.

Everyone agrees on two remarkable facts: The *Columbia* recovery was the largest ground search effort in American history; and it was also one with no internal strife, bickering, or inter-agency squabbles. Everyone involved had a single goal and worked collectively to achieve it—to bring *Columbia* and her crew home.

It was as true at KSC as it was in East Texas—acquaintances became friends; friends became good friends; and good friends became close for life. The experience changed the lives of the astronaut families and friends in unimaginable ways. It also changed the twenty-five-thousand-person team.

It made me a better launch director and, I'd like to think, a better person too. It made me appreciate it more when people sought me out for advice and guidance, so I opened up a little bit more. I had a much deeper appreciation for the astronauts and the risks of spaceflight. This served me well over the final twenty-two shuttle missions, by making me dig even deeper before giving that final "Go" on launch day. It became my habit to look out the windows of the Launch Control Center at the shuttle on the launchpad and think about my friends who were about to take an incredible risk. Was I truly ready to say,

"Go"? Was I ready to commit them to an ultimately risky endeavor? During this "gut check"—the ultimate gut check—Rick Husband and his crew, and the crew's families, were always with me in the decision. How could I ask another crew to go and *not* think about them? They were instinctively part of every launch decision.

The recovery of *Columbia*'s crew and the ship's debris changed the good people of East Texas as well. NASA was now part of their families, and they were part of the NASA family. Astronauts were no longer just faces on a TV screen. They were real people with whom the citizens had worked selflessly side by side—real men and women with families and loved ones who bore tremendous sacrifices so that America could accomplish its goals in space. And one of their beloved citizens and his pilot had given their lives to help NASA find *Columbia*.

They, and all the thousands of people from across the United States who participated in the recovery, need to know NASA will forever be in their debt and will always admire their selflessness.

The good people of *Columbia*'s recovery and reconstruction individually and collectively rose above all reasonable expectations. They succeeded in ways outwardly observable and only inwardly known.

All of them—all twenty-five thousand men, women, and children—are American heroes.

# EPILOGUE

I feel honored to be asked to write the epilogue for *Bringing* Columbia *Home*. This is a book that needed to be written, and I am so glad it is here. It tells the true story of an epic undertaking. Out of disaster came passion and meaning. People came together to search for *Columbia*, in part due to their sorrow for the loss of the crew, and in part due to their devotion to the mission of space exploration. This story also gives us a chance to reflect and to ask ourselves why we explore.

I would never want to relive that terrible Saturday morning of February 1, 2003. My crew and I were five weeks from launch and completely immersed in our own upcoming mission preparation when we lost *Columbia*. All of a sudden, our mission became displaced and distant. I immediately found myself thinking of each *Columbia* crew member, my last time spent with each of them, their families, and their future plans that would now never happen.

While the Space Shuttle Program immediately activated its contingency action plan, my crew focused on assisting the *Columbia* families, the accident investigation, and eventually the return-to-flight effort. We were called into entirely different roles than the ones for which we were training. Pilots and mission specialists became family assistance officers, Shuttle Program representatives, and even public spokespersons. I did not see my pilot Jim Kelly for three months while he served as the family assistance officer for Laurel Clark's family.

Steve Robinson likewise aided Mike Anderson's family. Steve eventually redesigned our patch to honor the STS-107 crew. Andy Thomas wrote a beautiful memorial to the *Columbia* crew, which was downlinked when our mission eventually flew. Our message was that we needed to continue *Columbia*'s mission of exploration.

The *Columbia* crew's mission gave them the chance to fulfill their dreams. It was an opportunity to fly to space, have an adventure, visit microgravity, live and work in a totally new and unusual environment, experience a feeling and freedom that you cannot possibly simulate on the surface, look down on our beautiful planet, push yourself to achieve the mission's challenging goals, and be part of a team with a meaningful vision.

That vision included understanding the universe we live in, the human body, the possibilities of new technologies, our Earth's natural processes, the secrets of our neighboring planets, and the mysteries of deep space. The *Columbia* crew was taking baby steps, but great missions begin with small steps—learning steps. They were passionate about their mission. Passion, like risk, is a part of any great quest.

I often tell my children that generations pass and centuries pass, but it seems that the sense of curiosity in people does not change. Sure, our environment changes and technology changes, but people are still human. We still carry the spirit and adventure of those we read about in history, the Bible, the Greek plays, the discoveries of Columbus, and the exploration of the Americas. Likewise, astronauts love to explore, and they feel very confident, focused, and determined about it. I believe exploring and taking risks will be around for a long time.

I would like to share a brief memory of each of the *Columbia* astronauts. Rick was a man that any astronaut would want as their commander. He was wise and charismatic. People often praised his beautiful singing voice. Willie was so proud of his children! I remember the day he stopped in my office to show me his son's artwork. Laurel and I had many conversations about the challenges of being both a mom and an astronaut. Mike was also a family man, and he always seemed to have a smile on his face, especially when he was going out to fly the T-38. Ilan often asked me to fly with him in the T-38. He was very humble and quiet, and clearly dedicated to his family. I wish I had gotten to know him better. KC and I worked together in the simulator, trying to solve the sneaky problems presented to us by the training team. She had such a calm and logical way about her. And Dave was an optimist. No matter how tough a problem he faced, Dave was a man of ideas and solutions.

It is still difficult for those of us who knew them to remember them now, as we miss their being physically here with us. But as time passes, I believe we should recall the happiness they brought to the people in their lives. We should celebrate their spirit and their love of space exploration. While I grieved over the loss of our friends, Evelyn Husband would remind me of a quote from Proverbs 3:5: "Trust in the Lord with all your heart and lean not on your own understanding."

My respect and thanks go out to those who designed, built, operated, and maintained the shuttle. Several technicians at Kennedy Space Center remarked to me that the loss of *Columbia* was in some ways like the loss of a family member. I know that they grieved her loss. *Columbia* was the flagship of the fleet. I had the privilege of commanding my third mission on *Columbia* in 1999 when my crew deployed the Chandra X-ray Observatory. So I completely understand it when I hear how the folks at KSC felt. They raised and nurtured *Columbia* throughout her lifetime.

The space shuttle was an amazing and versatile ship. Despite two tragedies, its successes were manifest. It was an engineering marvel. The shuttle was a test program, but it achieved its ultimate goal, which was to assemble the International Space Station. Other accomplishments of the shuttle include learning about the human body, in-space technologies, our planet, and deep space. We have inspired over thirty years of students to study math, science, and engineering. We worked very productively with people from many other countries, on many different levels. People of very diverse backgrounds came together on the shuttle.

The *Columbia* crew embodied the benefits of diversity. They were adventurers, explorers, and role models. How fortunate they were to have had the opportunity to experience a whole new world. Unfortunately, we will never be able to thank them for their contributions to the great journey of human discovery.

Of course, the *Columbia* and *Challenger* accidents have reminded us we need to be ever vigilant. Despite more than two years of careful work to prevent foam shedding from the shuttle's external tank, my STS-114 mission lost a large piece of foam on ascent, in a circumstance very similar to what happened to *Columbia* on STS-107. Preventing foam loss was a top objective of the return-to-flight effort, and while this turned out to be an embarrassment, I believe it sent a clear message—future boosters and spacecraft should be designed to protect the ship's reentry system (the heatshield), because rockets will always shed "stuff" like insulation and ice during the tumultuous minutes of ascent to orbit. This is why we will see future spacecraft designed with the reentry ship on top of the rocket, rather than beside it, as was the case with the space shuttle. The STS-114 incident was a very sobering reminder that a complex system like the shuttle can never be made completely safe, despite everyone's best efforts. Our future space travelers will be safer due to the lessons learned from the shuttle missions.

I must close by expressing my heartfelt thanks to the many thousands of people who worked tirelessly in East Texas. Because of their focus, passion, and long hours, we were able to recover enough of *Columbia* to determine the cause of the accident and return the shuttle to flight. I experienced their work firsthand when my crew visited the search areas several times, and we even assisted with a search. Soichi Noguchi from my crew found a piece of tile. We were proud to share in the efforts, but we primarily wanted to meet the people and get a feeling of what they were enduring in this difficult search. They worked under emotional stress, time constraints, cold winter weather, and changing scenarios. They completed their unprecedented mission—never had there been a search effort of this magnitude. Due to their efforts, the shuttle returned to flight, and the Space Station was completed.

So I say to all those who took part in the search effort: you too have made an important contribution to space exploration. I thank you personally, and I thank you on behalf of the Shuttle Program. Your story has now been told, and you can be proud of your role in the great journey of human discovery.

**Col. Eileen Collins, USAF, Retired**
Pilot, STS-63 and STS-84
Commander, STS-93 and STS-114

# AUTHORS' NOTES AND ACKNOWLEDGMENTS

February 1, 2018, marks the fifteenth anniversary of the *Columbia* accident. Merely counting the years is not in itself a sufficient justification for writing a book. However, the passage of a decade and a half does offer an opportunity to tell such an important story about American courage, compassion, and commitment while firsthand accounts are still available.

Because of the vast number of people involved in preparing *Columbia* for launch, searching for the crew and the debris of the vehicle, and reconstructing the shuttle, it was impossible for us to include all the stories of heartbreak, heroics, and selfless acts that we heard during our research for this book. Likewise, we deliberately pared back our technical discussions to make this book accessible to a broader audience. And we had to omit some material purely in the interest of producing a book that people could pick up without a forklift.

The authors maintain a blog at www.bringingcolumbiahome.com. Our regular posts supplement this book's chapters, and we include many more photographs than we could fit into the book. There are also links to related sites, videos, and other material to explore for people interested in *Columbia* and the Space Shuttle Program. The blog affords readers the opportunity to share their own memories of *Columbia* and to interact with the authors. We welcome your comments and hope that you will contribute to the recorded history of this remarkable time.

To the citizens of all of the communities throughout Texas and Louisiana where the *Columbia*'s debris came down: We hope that you will understand that space limitations and the need to tell a coherent story may have led us to say less about your particular county's contributions than we should have. Every person in Texas and Louisiana who was part of the recovery effort deserves to be proud of his or her accomplishment. Please know the NASA family is forever in your debt.

We recognize that government contractors did much of the heavy lifting in the recovery and reconstruction. For the sake of brevity, when we discuss a particular agency performing a function, we intend the reference to include both civil servants and the contractors who acted on behalf of that agency. Thus, a reference to "NASA personnel" may imply NASA employees as well as staff from United Space Alliance, Boeing, and other contractors who were working side by side with the civil servants. The transparent, "badgeless" environment that existed during this operation was one of the keys to its success.

Two fellows cannot write a book about twenty-five thousand Americans without a lot of help. We would like to take this opportunity to thank a number of people whose assistance and encouragement made this book possible.

First, we owe our collaboration on this book to the inimitable Norm Carlson, chief NASA test director during the first part of the shuttle era. Norm passed away on March 1, 2015. Mike and Jonathan met for the first time at Norm's memorial service. We decided to go to lunch together and get to know each other better. At that first lunch, Mike said, "I always threatened to write a book about *Columbia*." Jonathan said, "If you wanted to collaborate, I'd be glad to work with you." And the rest, as they say, is history. We're grateful to Norm, who brought us together, and whose memory constantly inspires us.

Greg Cohrs was as dedicated to helping this book succeed as he was to ensuring the success of the *Columbia* recovery effort. His technical assistance, his keen memory, his eye for detail, and his affable demeanor made him an incredible collaborator. We were proud to shine the spotlight on this wonderful man.

Marsha Cooper and Belinda Gay invited us to Hemphill and set up interviews with dozens of the people you have read about in this story. Their passion for keeping the memory of *Columbia* alive knows no bounds.

Jan Amen, Mark Stanford, Pam "Pambo" Melroy, Dom Gorie, Robert Hanley, Robert Pearlman, Michael Key, and Patrick Adkins provided us with detailed notes and with photographs that greatly enriched our understanding of the recovery and reconstruction efforts. Pambo, Jim Wetherbee, Steve Altemus, Mike Ciannilli, Dave Whittle, and Sean O'Keefe reviewed sections of the book for technical and historical accuracy. Astronauts Jerry Ross, Eileen Collins, John Herrington, and Bob Crippen also went far beyond the call of duty in assisting us during the research and writing process. Jeff Williams from Stephen F. Austin State University, who generated maps for searchers in February 2003, graciously produced several maps for this book using data from the search database.

To everyone we interviewed for the book (who we've listed in the "Interviewees" section), we thank you for sharing your memories, your pains and sorrows, and your moments of personal triumph. Your contributions made this book a story rather than a recitation of facts.

We thank our agent, the renowned military historian Jim Hornfischer, for his sage advice and for helping get our book in front of the Right Person. Maxim Brown, our editor at Skyhorse Publishing, was that Right Person. He made the publication process as painless as authors could ever wish it to be. We're grateful to him for giving us the latitude to write the book we wanted to write.

Scott Mack helped us structure the story, and he reviewed our drafts. How can you possibly go astray when you have a multi-time Teacher of the Year from the Fairfax County (Virginia) School System looking over your shoulder? Coralee Leon reviewed our draft manuscript with the keen eye of a highly successful editor and author. She made sure we were telling a coherent story. Thanks also to Penny Ward for transcribing many of the interviews and for reviewing the manuscript.

Susan Roy and Holly Williams acted as excellent coaches, spending countless hours talking Jonathan down off the ledge and lending their ears and encouragement when the going got tough. Jonathan's wife Jane gave him love, support, and space when he needed it most.

And to Charlotte, affected as much by the tragedy and its aftermath as Mike, you were and continue to be his perfect partner in all things.

# INTERVIEWEES

The following people graciously offered their time and thoughts during the research for this book. Most initial interviews were conducted between April 15, 2015, and September 30, 2016. Follow-up interviews were conducted as necessary to clarify and confirm information as the book developed.

| | |
|---|---|
| Adkins, Patrick | KSC quality assurance; identified and processed debris at the collection center in Hemphill |
| Alexander, Mike | Volunteer searcher, Sabine County |
| Allen, Mark | Logistics chief for the Sabine County Incident Command Center during the initial search; now is county judge of Jasper County. |
| Altemus, Steve | *Columbia* reconstruction director |
| Amen, Jan | Texas state fire chief's office; photographed much of the recovery activity in East Texas for the Texas Forest Service and NASA, escorted astronauts and crew families |
| Angermeier, Jeff | NASA debris recovery coordination from Lufkin |
| Arriëns, René | United Space Alliance closeout crew member in KSC's "White Room"; debris recovery operations in Hemphill |
| Awtonomow, Debbie | NASA manager of the debris collection and processing center at Hemphill beginning in late February |
| Bean, Olen | Branch fire coordinator, Texas Forest Service |
| Biegert, John | Crew module reconstruction lead engineer |

| | |
|---|---|
| Borsi, Mark | Head of security at KSC, ran security NASA security operations for Mishap Investigation Team out of Barksdale |
| Bridges, Roy | Astronaut; Kennedy Space Center director at the time of the accident |
| Cabana, Robert | Astronaut; director of Flight Crew Operations during the accident |
| Ciannilli, Mike | NASA test project engineer, air searcher during recovery operations; current curator of *Columbia* Preservation Office; director of NASA's Apollo 1/*Challenger*/*Columbia* Lessons Learned Program |
| Cohrs, Greg | US Forest Service timber sale forester in Sabine County National Forest; planned and directed crew recovery search operations in Sabine County; branch director for the remainder of the search in Sabine and San Augustine Counties |
| Collins, Eileen | Astronaut; commander of STS-114 "Return to Flight," the first mission after *Columbia* accident |
| Comer, Jim | Reconstruction lead engineer for United Space Alliance |
| Cooper, Marsha | US Forest Service staff in Sabine County; searcher, media relations during recovery; one of the driving forces behind the *Columbia* memorials in Sabine County |
| Cowart, Jon | Reconstruction lead engineer; JSC Orbiter Project Office |
| Crippen, Robert "Bob" | Astronaut; first pilot of *Columbia*; delivered eulogy for the ship and crew at the KSC runway after the accident |
| Eddings, Don | Texas Forest Service air search operations |

| | |
|---|---|
| Furr, Jim | Space Flight Awareness representative for NASA/United Space Alliance; coordinated morale events for searchers and community relations in Nacogdoches |
| Garan, Ron | Astronaut; first person to see the recovered video recorded by *Columbia*'s crew at the start of reentry |
| Gay, Belinda | Ran the volunteer feeding operation at Hemphill's VFW; one of the driving forces behind *Columbia* memorials in Sabine County |
| Gay, Roger | Commander, Hemphill VFW |
| Gehman, Admiral Harold "Hal," USN (ret) | Chairman, Columbia Accident Investigation Board (interviewed by email) |
| Gibson, Robert "Hoot" | Astronaut; commander of *Atlantis* STS-27 mission that returned from space heavily damaged by launch debris |
| Gorie, Dom | Astronaut; member of STS-107 Mishap Investigation Team; one of the "air bosses" for aerial debris searches |
| Gray, Mary Beth | Hemphill florist |
| Hamilton, Doug | US Forest Service law enforcement officer in Sabine County; one of the first responders on site after the accident |
| Hanley, Robert | JSC Flight Crew Operations, Melroy's deputy in crew module reconstruction |
| Herrington, John | Astronaut; one of the "air bosses" for aerial search for *Columbia* debris; first Native American astronaut |
| Holmes, Felix "Bubba" | US Forest Service heavy equipment operator, assisted with recovery of *Columbia*'s "nose cone" and the US Forest Service helicopter after the fatal accident |

| | |
|---|---|
| Iles, Don | City Manager, Hemphill |
| Ippolito, Gay | Public Affairs Officer for US Forest Service at Lufkin during the recovery |
| Jett, Brent | Astronaut; forward coordinator for crew recovery operations in Hemphill |
| Keifenheim, Jack | Lead for payload reconstruction |
| Kelly, Mark | Astronaut; first NASA person on site for a *Columbia* crew recovery on the day of the accident |
| Kelly, Scott | Astronaut; assisted with crew search starting Day 2 |
| King, Dave | Deputy director of Marshall Space Flight Center, headed overall NASA efforts for the crew and debris recovery from the Lufkin command center |
| Kovacs, Greg | Stanford professor of engineering and medical doctor; CAIB support researcher at the reconstruction hangar |
| Lane, Terry | FBI special field agent in East Texas; led FBI teams in crew recoveries in Sabine County |
| Leath, Beverly | Widow of Sabine County Judge Jack Leath |
| Leinbach, Charlotte | Mike's wife |
| Leinbach, Mike | KSC launch director; leader of Rapid Response Team to Barksdale on day of accident; leader of reconstruction effort |
| Maddox, Tom | Sabine County Sheriff; co-incident commander for the Hemphill Incident Command Team during the crew recovery |
| Mangiacapra, Amy | United Space Alliance staff in reconstruction hangar; helped set up the *Columbia* room in the VAB and managed it for 10 years |
| Mango, Ed | Assistant launch director; led the debris recovery effort in the field |

| | |
|---|---|
| McCowan, Hivie | Hemphill resident; volunteer food service worker nicknamed "Sweet Tea" at the VFW Hall |
| McLellan, Marty | Spacehab executive |
| Melroy, Pam | Astronaut; led the crew module reconstruction team and co-wrote the official crew survival report |
| Micklos, Ann | Thermal protection engineer, dated *Columbia* astronaut Dave Brown; worked in tile system reconstruction |
| Mills, Steve | High school principal, Hemphill |
| Millslagle, Jeff | FBI supervisor in Tyler, TX; managed the FBI resources during the recovery |
| Mott, Cecil Paul | City electrician for Hemphill |
| Moynihan, Linda | Administrative assistant for United Space Alliance; helped coordinate logistics in Barksdale and Texas for staff arriving from Kennedy Space Center |
| Nelson, Marie | Sabine County resident; volunteered at the Hemphill VFW |
| O'Keefe, Sean | NASA Administrator at the time of the accident |
| Ostarly, Larry | United Space Alliance chief for recovery operations from Barksdale and Lufkin |
| Raney, "Brother Fred" | Pastor of Hemphill First Baptist Church and head of Hemphill's Volunteer Fire Department; performed last rites at each crew recovery location |
| Readdy, Bill | Former astronaut; NASA associate administrator for manned spaceflight at the time of the accident |
| Rhode, Linda "Agent 99" | KSC security special agent |
| Riley, Dwight "Grandpa" | Sabine County resident; volunteer searcher |

| | |
|---|---|
| Ross, Jerry | Astronaut; head of Flight Crew Operations support at KSC; set up processes for debris recovery in Texas along with Ed Mango; lead astronaut during the debris recovery effort |
| Ross, Renée | Crew module reconstruction support for on-board document identification |
| Sauerwein, Dan | Worked at NASA/Neutral Buoyancy Lab in Houston; spent a week as a volunteer searcher in Hemphill |
| Scales, Tommy | State trooper with Texas Department of Public Safety |
| Schumann, Gerald (Gerry) | Director of safety at KSC; on the ground in Hemphill during the debris recovery |
| Smith, Billy Ted | Emergency management coordinator for East Texas Mutual Aid Association; co-incident commander (with Sheriff Maddox) in Sabine County during the crew search operations |
| Smith, Pat | Bank officer in Hemphill at the time of the accident |
| Sowell, Jamie | US Forest Service forestry technician, led a volunteer search team in Sabine County during crew recovery operations |
| Stanford, Mark | Texas Forest Service overall lead in the recovery operations |
| Starr, Byron | Hemphill funeral director (Squeaky's son); also on site for most recoveries; wrote a book about the crew recovery |
| Starr, John "Squeaky" | Hemphill funeral director; was on site for most of the crew recoveries and transported crew remains |
| Stilson, Stephanie | NASA debris recovery site manager in Nacogdoches |
| Thurston, Scott | *Columbia* processing flow manager; led *Columbia* preservation team to determine best |

| | |
|---|---|
| | approach for preserving and learning from the ship's debris |
| Walker, Charles "Boo" | Texas Forest Service air operations; air-traffic controller for the helicopter searches |
| Wells, Scott | FEMA federal coordinating officer; led FEMA's *Columbia* recovery operations in Texas |
| Wetherbee, Jim | Astronaut; directed operations for the search and recovery of *Columbia*'s crew |
| Whittington, Sunny | Hemphill elementary school teacher who had her class make sandwiches for searchers |
| Whittle, Dave | Director of NASA's Mishap Investigation Team at Barksdale and Lufkin |
| Williams, Jeffrey | Geographic Information System (GIS) first responder, Stephen F. Austin State University, Nacogdoches |
| Willoughby, Jeremy | Florida wildland firefighter who was on the crew that found the OEX recorder in San Augustine County |

# ACRONYMS AND TECHNICAL TERMS

All-risk incident (also called "all-hazard incident")—any incident or event—natural or human-caused—warranting action to protect life, property, environment, and public health and safety, and minimize disruption of governmental, social, and economic activities.

Apollo 1—NASA's first fatal spacecraft accident. The crew of three astronauts was killed in a fire during a launchpad test on January 27, 1967.

Astronaut—anyone who has flown in outer space, defined as starting at 100 km altitude.

Bipod—a two-legged strut that attached the nose of the space shuttle orbiter to its external fuel tank during the ride to orbit. Explosive bolts severed the bipod's connection to the orbiter when the vehicle reached orbit.

Bipod ramp—an aerodynamic wedge of foam insulation about 30 inches long, 14 inches wide, and 12 inches tall, which covered the fittings that attached a bipod strut to the external tank. The insulation was intended to keep ice from forming on the fittings, which could have been dangerous to the shuttle if the ice was dislodged during ascent. After the *Columbia* accident, heaters were added to the bipod fittings, eliminating the need for foam.

CAIB—see *Columbia* Accident Investigation Board.

Cape Crusader—an astronaut who supports the flight crew and processing activities at Kennedy Space Center during preparation for and at the end of a mission.

*Challenger*—the second orbiter to fly in space. The vehicle was destroyed during ascent in an accident on January 28, 1986, killing its crew of seven astronauts.

*Columbia*—the first orbiter to fly in space. Its first mission launched April 12, 1981. *Columbia* was destroyed during reentry over Texas on February 1, 2003, killing its crew of seven astronauts.

*Columbia* Accident Investigation Board (CAIB, pronounced "kabe")—an independent investigation board tasked with determining the cause of the *Columbia* accident and making recommendations regarding changes to organizations, practices, policies, procedures, and flight hardware to prevent future accidents.

Crew module—the orbiter's crew compartment, consisting of the flight deck (cockpit), mid-deck, and avionics bays. The crew rode into and back from orbit within the crew module, which also contained their toilet, sleeping berths, galley and food, and storage lockers.

Cryos (short for "cryogenic gases")—gases that have been supercooled to the point that they have become liquefied. The shuttle's main engines burned liquid hydrogen as fuel (stored at minus 423°F) with liquid oxygen (stored at minus 297°F) as the oxidizer. Both propellants were carried in the shuttle's external tank. The orbiter itself carried smaller tanks of liquid helium to pressurize the maneuvering system propellant system and liquid oxygen and liquid hydrogen that were used by the shuttle's electricity-generating fuel cells.

DPS—Texas Department of Public Safety.

Elevon—a control surface combining the functions of an elevator and aileron. There were two elevons (inboard and outboard) on the back end of each wing on the shuttle.

EPA—the US Environmental Protection Agency.

External tank (ET)—the only nonreusable part of the shuttle. The external tank contained the liquid hydrogen and liquid oxygen powering the space shuttle's main engines during ascent to orbit. The tank was covered by a layer of orange-brown, sprayed-on insulating foam, which kept ice from forming on the tank.

Extra-vehicular Activity (EVA)—a "space walk," in which an astronaut leaves the protective confines of the spacecraft to work outside in the vacuum of space.

FEMA—the Federal Emergency Management Agency.

Fire crews—teams of twenty men and women contracted by the US Forest Service to fight wildfires and respond to other natural and human-caused incidents.

Firing Room—one of four control rooms in Kennedy Space Center's Launch Control Center in which the launch team ran the testing, countdown, and launch of a space shuttle mission.

Flight Director—the leader of the flight control team in the Mission Control Center, with overall responsibility for mission operations and decisions regarding safe flight. The flight director assumed control of a mission upon ignition of the solid rocket boosters at T-0.

Grid search—a technique for systematically searching an area. Searchers space out abreast at a prescribed interval and walk forward together, each person scanning the ground around them. When they reach the assigned distance from the starting point, the line pivots around one of the end people, and the group walks in the other direction. Grid searching increases the likelihood of finding a missing person or object, since it ensures that every square foot of the search area is covered.

Heatshield—the protection system around a spacecraft that helps it survive the heat of reentering the Earth's atmosphere. The heatshield system of the space shuttle included silica tiles on its underside, reinforced carbon-carbon panels on its nose and wing leading edges, and quilted silica and felt blankets on areas not subject to high heating.

Hypergolic propellants (also called "hypers")—chemical liquids used to power the orbiter's onboard maneuvering system rocket engines and thrusters. Hypergolic chemicals ignite instantly when the fuel and oxidizer come in contact with each other. The shuttle used monomethylhydrazine as fuel and nitrogen tetroxide as the oxidizer. Both chemicals are extremely toxic, both through inhalation and contact with skin, and cause death with even very limited exposure. Tanks of both propellants were located in various places throughout the orbiter.

Incident command system (ICS)—a standardized process for command, control, and coordination of multiple agencies in response to an emergency situation.

Incident Management Team (IMT)—a group that responds to an emergency using the incident command system framework. IMTs are "typed" according to the scope and complexity of the incidents they are certified to manage.

Inconel—an alloy of nickel, chromium, iron, and other metals used for spacecraft parts subject to high-temperature and/or high stress environments.

International Space Station (ISS)—a microgravity and space environment laboratory in low Earth orbit, funded and manned by the United States, Russia, Japan, the European Space Agency, and Canada. It has been continuously occupied since November 2, 2000, recently by six astronauts at a time. Major assembly of the ISS occurred between 1998 and 2011 with unmanned vehicles delivering the Russian segment modules and the space shuttle delivering the components for the US side of the ISS.

Johnson Space Center (JSC)—NASA's center in Houston, Texas, which is home to the astronaut corps, the engineers who designed the space shuttle, the Space Shuttle Program Office, training facilities, and the Mission Control Center. Often referred to simply as "Houston."

Kennedy Space Center (KSC)—NASA's principal launch facility located on the Atlantic coast of central Florida, which prepares and launches manned and unmanned spaceflight missions.

Launch Control Center (LCC)—a four-story building at Kennedy Space Center that serves as the hub for conducting NASA test and launch operations. The LCC has four Firing Rooms, computer and communications systems support areas, as well as offices and conference rooms.

Launch Director—the head of the launch team at Kennedy Space Center, responsible for making the final "Go for launch" decision.

MADS recorder—see OEX.

Marshall Space Flight Center (MSFC)—NASA's center in Huntsville, Alabama, which designs the propulsion systems for NASA's manned spacecraft.

Michoud Assembly Facility—NASA's facility near New Orleans, Louisiana, where the space shuttle's external tanks were built and tested. The tanks were shipped by barge from Michoud to Kennedy Space Center.

Mishap Investigation Team—a multidisciplinary NASA internal team responsible for debris recovery, protection, and impoundment immediately after a spacecraft accident.

Mission Control Center—the building at Johnson Space Center housing the flight control room where flight directors and flight controllers managed a space shuttle mission from launch until landing.

Mission Management Team (MMT)—a group of managers from all aspects of the Shuttle Program throughout NASA and its shuttle contractors. The MMT held reviews to clear a shuttle mission for launch and was supposed to meet every day during a mission to keep leaders informed and, if necessary, debate risks and solutions to issues occurring before or during the flight, especially those outside documented launch or flight procedures.

NTSB—the National Transportation Safety Board.

OEX—the Orbiter Experiments recorder, akin to an airplane's flight-data recorder, which recorded on magnetic tape the state of hundreds of pressure, temperature, motion, and other sensors inside *Columbia*. Also called the Modular Auxiliary Data System (MADS) recorder.

OMS pods (pronounced "ohms")—removable flight structures on each side of the aft end of the orbiter's body, at the base of the vertical stabilizer (tail), containing the Orbital Maneuvering System (OMS) engines and maneuvering thrusters.

Orbiter—the winged vehicle at the heart of the Space Transportation System. The orbiter was about the same size as a DC-9 or MD-80 commercial airliner. It took off like a rocket and landed like a glider.

Orbiter Processing Facility—one of three specially outfitted hangars near Kennedy Space Center's Vehicle Assembly Building in which orbiters were maintained, repaired, and configured for upcoming missions.

Payload bay—the cargo compartment in the middle section of the space shuttle. The payload bay was fifteen feet in diameter and sixty feet long—about the size of a school bus. The shuttle could carry up to sixty thousand pounds of payload into Earth orbit.

Plasma—the "fourth state of matter," a superheated gas with approximately equal numbers of positively charged ions and electrons, created during reentry as a fast-moving spacecraft compresses and superheats the surrounding atmosphere.

Pyrotechnic devices ("pyros")—small explosive charges for initiating dozens of critical actions on the space shuttle. Some of the functions performed by pyros included severing the bolts that held the shuttle to the launch platform at the moment the solid rocket boosters ignited, separating the external fuel tank from the orbiter, deploying the drag chute when the orbiter landed, blowing the hatch for emergency egress—even lowering the landing gear.

Rapid Response Team (RRT)—a team of about ninety engineers and technicians from Kennedy Space Center who deployed if the shuttle made a "nonroutine" landing, tasked with securing the vehicle and bringing it back to KSC.

Reinforced carbon-carbon (RCC)—also known as carbon-fiber-reinforced carbon. A composite material made up of layers of rayon cloth impregnated with a carbon resin and then baked into a hard surface. It is used for structural applications in situations subject to extremely high temperatures, such as the space shuttle's nose and leading edge of its wings, the nose cones of intercontinental ballistic missiles, and Formula One car disc brakes. RCC is structurally strong and tough, but brittle when impacted with sufficient force.

Shuttle Landing Facility (SLF)—the concrete runway at Kennedy Space Center where the space shuttle returned at the end of its missions. At 15,000 feet in length, the SLF is one of the longest runways in the world.

Solid rocket booster (SRB)—one of two large solid-propellant motors that together provided 83 percent of the thrust in the first two minutes of the space shuttle's flight. SRBs separated from the external tank and parachuted into the ocean, to be reused on later missions.

Soyuz—the Russian manned spacecraft used to ferry crews of up to three people to and from low Earth orbit.

Spacehab—a pressurized module carried in the shuttle's payload bay and connected to the crew module's air lock by a tunnel. Spacehab modules carried scientific and medical experiments the crew could operate in a shirtsleeve environment while on orbit.

Space shuttle—see Orbiter.

STS-xxx—abbreviation for "Space Transportation System." NASA designated space shuttle missions as STS followed by a number. The numbering was sequential based on the order in which the flights were initially approved. Priority changes or equipment problems occasionally caused some missions to be moved ahead or back in the launch manifest. *Columbia*'s final flight was STS-107, but it was the 113th shuttle mission to fly.

T-0 (pronounced tee zero)—the moment in the countdown when the shuttle's solid rocket boosters ignite, explosive holddown bolts sever, and the shuttle lifts off the launchpad.

T-38—two-seat, supersonic jet aircraft used by the astronauts for training, transportation, and maintaining flying proficiency.

Terminal Countdown Demonstration Test (TCDT)—the unfueled dress rehearsal for a space shuttle countdown and launch. The crew was strapped into their seats in the orbiter while the launch team supported them from the Firing Room.

Thermal protection system (TPS)—see Heatshield.

Tiles—blocks of low-density, porous silica bonded to the skin of the space shuttle to protect it from the heat of reentry. More than twenty thousand tiles—each with a unique shape—were on each space shuttle. They ranged from one to five inches in thickness depending on their location on the shuttle.

Trans-Atlantic Landing (TAL, pronounced "tal")—one of the abort modes available to the space shuttle if an emergency situation prevented it from going into orbit. The shuttle could land at designated landing sites in West Africa and Europe, as well as several air bases along the US East Coast.

Type 1 Team—the most highly trained and qualified level of Incident Management Team, certified to work in response to complex national and state level emergency situations.

United Space Alliance (USA)—the contractor responsible for most space shuttle operations and maintenance at KSC. In the mid-2000s, about 13,000 government and contractor employees were at KSC; 8,100 of those were USA staff.

Vehicle Assembly Building (VAB)—the building at KSC where space shuttles were "stacked"—raised to a vertical position and mated to the external tank and solid rocket boosters in one of the four cavernous "High Bay" assembly areas before transport to the launchpad.

White Room—a small area on the launch tower from which an astronaut crew enters their spacecraft.

# NOTES

Except where otherwise noted, quotes and observations attributed to people throughout the book are taken from the authors' interviews with those individuals. Other detailed information that is not footnoted is from Mike Leinbach's personal notes and files.

**Chapter 2: Good Things Come to People Who Wait**

1. *Challenger* carried a crew of eight astronauts on the October/November 1985 flight of STS-61A. *Atlantis* also carried eight astronauts and cosmonauts on its return from the Russian *Mir* space station on STS-71 in 1995. Throughout the rest of the Space Shuttle Program, crew complement was limited to seven astronauts.
2. Tariq Malik, "NASA's Space Shuttle By the Numbers: 30 Years of a Spaceflight Icon," Space.com, March, 9, 2011, www.space.com/12376-nasa-space-shuttle-program-facts-statistics.html.
3. Shuttle mission numbers were assigned based on their original order in the launch manifest. STS-107 was the 107th assigned flight in the Space Shuttle Program. Changes in launch priorities and availability of hardware occasionally changed the order in which the missions flew. STS-107 was the 113th shuttle mission to fly. In informal conversation, missions were usually just referred to by their number—in this case, "one-oh-seven."
4. NASA, "Sixteen Minutes from Home: A Tribute to the Crew of STS-107," KSC Web Studio (Kennedy Space Center, FL) video, February 2003.

5. Each one of the four orbiters was in a constant state of flux—undergoing maintenance, being "de-configured" after returning from a mission, being prepared for an upcoming mission, being tested, being stacked, sitting at the launchpad, flying a mission, or being upgraded for safety and/or performance reasons. There were three hangars and four orbiters, occasionally requiring one of the vehicles to sit in an empty bay in the Vehicle Assembly Building while awaiting space in a hangar.
6. The STS-107 crew also ran an interface test in June 2001, before the STS-109 mission was moved ahead of STS-107 in the launch schedule. STS-107's experiments were removed from the payload bay, replaced with Hubble Space Telescope servicing equipment, and then placed back into the payload bay after *Columbia* returned from the STS-109 flight.
7. *Columbia*'s refit would have involved installing an air lock in the payload bay (*Columbia* was the only shuttle with its air lock inside the crew compartment) and removing much of the test instrumentation that added weight to the vehicle. This would have included removing the Orbiter Experiments (OEX) recorder, which was to prove crucial to investigating the accident.

## Chapter 3: The Foam Strike

1. *Challenger* was lost due to a cascading series of events that started with the failure of rubber O-rings in a joint of a solid rocket booster. Low-level engineers were unable to persuade middle management to delay the launch or even to seriously consider their concerns that the O-rings and the booster design might not work as intended in the frigid temperatures on the morning of *Challenger*'s January 28, 1986, launch day. Because middle management quashed the concern, the launch team and mission controllers were unaware that there was a potentially serious problem.
2. *Atlantis* on STS-27 holds the distinction of being the most heavily damaged spacecraft to return safely from orbit. The cork tip of the right-hand SRB fell off during ascent and gouged the side of the orbiter. More than seven hundred tiles were damaged, and one was knocked off completely. During reentry, plasma completely melted through a steel antenna cover under the missing tile and had started

melting the skin of the orbiter, but *Atlantis* passed through the period of peak heating before its airframe was breached.

3. Steve Stitch, email message to Rick Husband and Willie McCool, subject "INFO: Possible PAO Event Question," January 23, 2003, www.jsc.nasa.gov/news/columbia/107_emails/foamemails.doc.
4. Evelyn Husband and Donna VanLiere, *High Calling: The Courageous Life and Faith of Space Shuttle Commander Rick Husband* (Nashville, TN: Thomas Nelson, 2003), 163.
5. A study managed by LeRoy Cain subsequently determined that even if it were possible to jettison the payload bay contents and unneeded consumables on the ship, the maximum temperature reduction on the wing leading edges was at best only 7 percent ("Entry Options Tiger Team," NASA Mission Operations Directorate, Flight Director Office, April 22, 2003).

**Chapter 4: Landing Day**

1. NASA, *STS-107 Shuttle Press Kit* (Houston, TX: NASA, December 16, 2002), 17.
2. NASA added the Inflight Crew Escape System to space shuttles after the *Challenger* accident. Starting when the shuttle was at about thirty thousand feet in altitude, the astronauts could depressurize the crew module, jettison the hatch, and then extend an escape pole through the opening in the fuselage. One by one, the astronauts would hang a strap on the pole and slide out through the hatch. Their individual parachutes would open once they were clear of the shuttle. It took about ninety seconds to get the crew out this way, by which time the orbiter was at an altitude of ten thousand feet ("Everybody Out!" NASA Educator Feature, https://www.nasa.gov/audience/foreducators/k-4/features/F_Everybody_Out.html). This process would occur over the ocean, where the shuttle would be ditched.
3. NASA, *Columbia Crew Survival Investigation Report SP-2008-565* (Houston, TX: NASA Johnson Space Center, 2008), 1–29.

**Chapter 5: Recovery Day 1**

1. John Tribe, "Forever Remembered," unpublished personal recollections on the STS-107 mission.
2. Kim Anderson email, February 4, 2003, accessed at www.ars-fla.com/Mainpages/Spaceflight/STS-107/KimAnderson-STS-107.htm.

3. Within hours of assuming the role of NASA administrator in December 2001, O'Keefe asked his senior leaders to brief him on what would happen if something were to go wrong. NASA formalized a new contingency plan in November 2002, and revised it in January 2003, less than a month before the *Columbia* accident. O'Keefe later said in an interview with Jonathan Ward, "When you're in a situation like that, I don't know many people that can stand there and be as ice-cold as it takes to actually think through the proper series of those kinds of events. You can tell the difference between those who have an organized way of proceeding, versus those that say, 'What in the hell do we do now?' The one thing I didn't have to worry about on February 1, 2003, was what we were going to do next. The guy standing next to me on the runway has the binder that says, "Here's the plan, and here's how we do it, starting with: 'Item 1. Here's who we need to call and here's how we're going to set up a mishap investigation.'"
4. NASA, *NASA Accident Investigation Team Final Report* (Washington, DC: NASA Headquarters, August 22, 2003), iv. The fourteen organizations included the Columbia Task Force, the Headquarters Contingency Action Team, the NASA Accident Investigation Team, the Mishap Investigation Team, the Reconstruction Team, the Orbiter Vehicle Engineering Working Group, the Emergency Operations Center, the Data and Records Handling Working Group, the Early Sightings Assessment Team, the Systems Integration Working Group, the External Tank Working Group, the Space Shuttle Main Engine Working Group, the Reusable Solid Rocket Motor Working Group, and the Solid Rocket Booster Working Group.
5. "Grand Jury Indicts Man for Stealing Shuttle Toilet," *Lufkin Daily News* (Lufkin, TX), May 8, 2003 (www.lufkindailynews.com/news/newsfd/auto/feed/news/2003/05/08/1052368037.00303.3371.4691.html). The tank was recovered on May 7 in a strange turn of events, and its alleged thief became the fifth person to be indicted in East Texas for stealing debris from the shuttle. Eyewitnesses to his arrest stated that, by day, he was involved in the search for crew remains. He then allegedly went back to bring debris home from the field at night, hiding items under clothing in his trailer. He also had a stash of pyrotechnic devices—he did not know what they were or how dangerous they were—under his bed. Rumor held that he had a romantic encounter with a female searcher, and when his wife learned of the affair, she

notified local authorities of the stolen material in their home. Accused of withholding the "compactor tank assembly" from NASA's recovery efforts, he was also indicted by a grand jury on a charge of being a felon in possession of a firearm. In August 2003, he was acquitted of the theft charge in exchange for pleading guilty to the firearms charge.

6. *NASA Accident Investigation Team Final Report*, 57. The ten members assigned to the MIT for a mission were published via internal memo six weeks before flight.
7. Dave Whittle, "Columbia Accident Investigation Board (CAIB) Process Lessons Learned Video Interview" (Washington, DC: NASA Headquarters, 2013), www.nasa.gov/externalflash/CAIB/transcripts/whittle/whittle03.pdf.
8. Jan Amen email, February 4, 2003.
9. Donna M. Shafer and Amy Voigt LeConey, "First Hand Account of Selected Legal Issues from the Recovery and Investigation of the Space Shuttle Columbia," *Journal of Space Law*, vol. 30 (2004), 40, www.nasa.gov/externalflash/CAIB/docs/CAIB%20Law%20Review%20Article.pdf.
10. Phillip Stepaniak, executive ed., *Loss of Signal: Aeromedical Lessons Learned from the STS-107 Columbia Space Shuttle Mishap*, SP-2014-616 (Washington, DC: NASA, 2014), 33.
11. Shafer and LeConey, "Legal Issues," 42.
12. NASA, *Space Shuttle Columbia Material Recovery, Report CB-QMS-024* (Houston, TX: NASA Johnson Space Center Flight Crew Operations Directorate, September, 2004, unpublished), 2.
13. NASA, *Report CB-QMS-024*, 2.
14. Out of respect for the crew's families, NASA has never released details about the identity, location, or condition of any crew member's remains during the recovery.
15. Greg Cohrs, "Hemphill Recovery of the STS-107 Columbia, Notes of Greg Cohrs, May 28 through June 16, 2003," unpublished, 4.
16. Michael Cabbage and Robyn Suriano, "Fatal Return: A Stunned NASA Searches for Answers," *Orlando Sentinel* (Orlando, FL), February 2, 2003, 1–20.
17. Greg Cohrs email to Jonathan Ward, September 27, 2016.
18. Federal Emergency Management Association (FEMA), "FEMA Emergency Operations Vehicle (EOV)," fact sheet, undated [2003 or earlier].

19. Stepaniak, *Loss of Signal*, 34.
20. Cohrs, "Notes," 4.
21. Jeff Williams, interviewed by Connie Hodges, Center for Regional Heritage, Stephen F. Austin State University, March 24, 2003, digital.sfasu.edu/cdm/search/collection/col/searchterm/audio/field/title/mode/all/conn/and/order/nosort.

**Chapter 6: Assessing the Situation**

1. Greg Cohrs email to Jonathan Ward, December 19, 2016.
2. FEMA, "FEMA Puts Federal Resources into Action to Assist State and Local Authorities in Search, Find and Secure Mission for Columbia Debris," news release HQ-03-029, February 2, 2003.
3. Cohrs, "Notes," 5.
4. Cohrs, "Notes," 5.
5. Dom Gorie began alternating in this role with Horowitz several days into the recovery period.
6. US Navy, *US Navy Salvage Report, Space Shuttle Columbia*, Report S0300-B5-RPT-01 (Washington, DC: US Navy, Naval Sea Systems Command, September 2003), 1-7.
7. Stepaniak, *Loss of Signal*, 21.
8. Paul Keller, writer-editor, *Searching for and Recovering the Space Shuttle Columbia: Documenting the USDA Forest Service Role in This Unprecedented 'All-Risk' Incident, February 1 through May 10, 2003,* www.fireleadership.gov/toolbox/lead_in_cinema_library/downloads/challenges/Searching_Recovering_Shuttle_Columbia_2003_Paul%20Keller.pdf.
9. Byron Starr, *Finding Heroes: The Search for Columbia's Astronauts* (Vancouver, Canada: Liaison Press, 2006), 50.
10. NASA, "NASA Asks for Help with Columbia Investigation," news release H03-033, February 2, 2003.
11. FEMA, "FEMA Puts Federal Resources into Action to Assist State and Local Authorities in Search, Find and Secure Mission for Columbia Debris," news release HQ-03-029, February 2, 2003. This was a suite of Airborne Spectral Photo-imaging of Environmental Contaminants Technology (ASPECT) sensors mounted in a twin-engine aircraft.
12. Jan Amen email, February 4, 2003.
13. Stepaniak, *Loss of Signal*, 21.
14. Greg Cohrs email to Jonathan Ward.

**Chapter 7: Searching for the Crew**

1. Stepaniak, *Loss of Signal*, 25–7.
2. Stepaniak, *Loss of Signal*, 25–7.
3. Cohrs, "Notes," 6.
4. Starr, *Finding Heroes*, 81–8.
5. Interview with Marsha Cooper. Several years after the accident, the sister of one of *Columbia*'s crewmen came to Sabine County to visit the location where her brother had been recovered. Marsha Cooper was her escort and host. As they sat outside and talked the evening she arrived, the astronaut's sister told stories about her brother's childhood. She said that he used to enjoy fishing with their father, who would often remark about seeing a reflection of a white dog in the water. Cooper said she was stunned. This astronaut's remains were found near the water. His was the recovery at which the white dog had followed the sheriff and the rest of the group into the woods to the site. Cooper told the astronaut's sister about the incident, and they both broke into tears.
6. Interviews with Billy Ted Smith and Mark Allen; Sabine County incident management team press briefing on February 3, 2003, edition. cnn.com/TRANSCRIPTS/0302/03/ip.00.html.
7. Starr, *Finding Heroes*, 90.
8. Cohrs, "Notes," 6; interviews with Billy Ted Smith and Mark Allen.
9. Starr, *Finding Heroes*, 116.
10. Stepaniak, *Loss of Signal*, 23.
11. Starr, *Finding Heroes*, 91.
12. FEMA, "FEMA Updates Search, Find And Secure Activities For Columbia Emergency [4:00 p.m. Release]," news release HQ-03-035, February 4, 2003.
13. Cohrs, "Notes," 6.
14. Pete Churlon, "Space Shuttle Columbia Tragedy Photo Gallery," *Beaumont Enterprise* (Beaumont, TX), January 28, 2011, www.beaumontenterprise.com/photos/article/photo-548688; Pat Oden emails reprinted on www.hemphilltexas.com.
15. NASA, "Johnson Space Center Memorial Time Updated," news release H03-042, February 3, 2003.
16. NASA, "NASA Provides Update About Columbia Investigation," news release H03-051, February 4, 2003.

17. NASA, "Columbia Investigation."
18. NASA, "Space Shuttle Accident Investigation Board Chair Tours Recovery Area," news release H03-047, February 4, 2003.
19. Jan Amen email, February 4, 2003.
20. Interview with Jim Wetherbee; Stepaniak, *Loss of Signal*, 78.
21. *Loss of Signal*, 28–9.
22. FEMA, "FEMA Updates Search, Find And Secure Activities For Columbia Emergency [4:00 p.m. Release]," news release 3171-09, February 6, 2003.
23. James Hull email, February 7, 2003.
24. FEMA news release 3171-09.
25. "Today, Deputy NASA Administrator Frederick Gregory will render honors to the crew of the Space Shuttle Columbia at Dover Air Force Base in Delaware. *The remains of the orbiter's seven astronauts* are scheduled to arrive in flag-draped caskets at Dover about 2 p.m. EST on board a C-141 Starlifter. . . . The Charles C. Carson Center for Mortuary Affairs at the base will prepare the remains for return to the families. Ramon's remains will be flown to his home in Israel for burial," (NASA, "Deputy Administrator Meets Space Shuttle Columbia Astronauts' Remains at Dover AFB," news release H03-053, February 5, 2003, emphasis added).
26. Cohrs, "Notes," 8.
27. Starr, *Finding Heroes*, 182.
28. Interviews with Jerry Ross and Jim Wetherbee; ESRI, "Space Shuttle Columbia Debris Recovery Enhanced with GIS," Summer 2003, www.esri.com/news/arcnews/summer03articles/space-shuttle.html.
29. FEMA, "FEMA Updates Search, Find and Secure Activities for Columbia Investigation," news release 3171-13, February 7, 2003.
30. Cohrs, "Notes," 9.
31. "In Honor of the Columbia Shuttle Astronauts," Lufkin, TX, First Baptist Church, February 8, 2003.
32. Pat Oden email, February 8, 2003.
33. This motto appears on a commemorative T-shirt that Belinda Gay was wearing in a photograph dated February 10, 2003.
34. Interviews with Greg Cohrs, Terry Lane, Tom Maddox.
35. Stepaniak, *Loss of Signal*, 28.
36. Cohrs, "Notes," 13.

## Chapter 8: *Columbia* Is Going Home in a Coffin

1. FEMA, "FEMA Establishes Joint Information Center For Columbia Debris Search, Find, And Secure Mission At Lufkin Civic Center," news release HQ-03-031, February 3, 2003.
2. FEMA, "FEMA Continues to Coordinate Actions to Assist State and Local Authorities in Search, Find and Secure Mission for Columbia Debris," news release HQ-03-030, February 3, 2003.
3. FEMA, "FEMA Updates Search, Find and Secure Activities for Columbia Emergency [4:00 p.m. Release]," news release HQ-03-032, February 3, 2003.
4. US Navy, *Salvage Report*, 1–7.
5. Interviews with Dave Whittle and Larry Ostarly; Jim Wetherbee email to Jonathan Ward.
6. Pete Churlon, "Space Shuttle Columbia Tragedy Photo Gallery," *Beaumont Enterprise* (Beaumont, TX), January 28, 2011, www.beaumontenterprise.com/photos/article/photo-548675.
7. Starr, *Finding Heroes*, 69–72.
8. Pat Adkins email to Jonathan Ward.
9. Pat Adkins email to Jonathan Ward. Jerry Ross said that he had advocated for having all crew personal effects sent directly to the Astronaut Office in Houston. However, instructions were that everything recovered would be processed through the reconstruction hangar at Kennedy first, and then crew items would be sent to Houston.
10. FEMA, "FEMA Updates Search, Find and Secure Activities for Columbia Emergency [11:00 a.m. Release]," news release HQ-03-034, February 4, 2003.
11. FEMA, "FEMA Updates Search, Find and Secure Activities for Columbia Emergency [4:00 p.m. Release]," news release HQ-03-035, February 4, 2003.
12. FEMA news release HQ-03-034.
13. Shafer and LeConey, "Legal Issues," 61.
14. FEMA news release HQ-03-035.
15. Ener later achieved notoriety for entertaining NASA workers with his intricately fabricated tall tales of "one-armed space monkeys" that had escaped from *Columbia* and were sighted running loose in the woods of Sabine County. At one point, Ener even took out an ad in the local paper seeking to purchase monkey traps.
16. Starr, *Finding Heroes*, 69–72.

17. Crookshanks and Benzon, "CAIB Lessons Learned Video Interview."
18. Wetherbee noted in an email to Jonathan Ward that in the crew search operations, each error was corrected expeditiously based on the extensive experience and professionalism of FBI special agent Mike Sutton. Using Sutton's detailed and extensive system for logging the reported data, the crew remains leadership team in the Lufkin command center was able to rectify all errors.
19. ESRI, "Recovery Enhanced with GIS." The GIS and the EPA databases tracked different information. FEMA's new Shuttle Interagency Debris Database (SIDD) tracked all reports. EPA was only tracking the items that had actually been recovered. FEMA and EPA resolved the situation by agreeing that reports would first be entered into SIDD. Once it was clear that the report was for a new item rather than for one that had already been entered, the information would be sent to EPA's database. After EPA had investigated the sighting and collected an item, the information then went back to SIDD. The incident commanders could therefore use the data in SIDD to target search operations.
20. Interviews with Dave Whittle, Ed Mango, Jim Wetherbee; ESRI, "Recovery Enhanced with GIS"; *NASA Accident Investigation Team Final Report*, 60.
21. Greg Cohrs email to Jonathan Ward. Cohrs noted that one day later in the search, a military ordnance disposal team "cleared" one item in the field as nonexplosive, which the search team interpreted as meaning "safe." Cohrs said, "It was actually high-pressure, and we were at risk moving it by hand and vehicle to the collection center, as was pointed out when I delivered it. It was later depressurized on a shooting range. We made our personnel aware of that type of hazard."
22. FEMA, "Substantial New Resources Committed to Expedite Search and Collection Effort for Columbia Material," news release HQ-03-036, February 5, 2003; FEMA, "Columbia Material Collection Guidelines: Fact Sheet," news release HQ-03-036a, February 5, 2003. Shuttle radios used classified, military-grade communications security technology to prevent unauthorized access.
23. FEMA, "Seven West Texas Counties Alerted of Possible Scattered Shuttle Material," news release 3171-12, February 6, 2003.
24. Five Texas residents were charged with stealing debris from the shuttle. None ever served time for the thefts. Jeffrey Gettleman, "Loss of the

Shuttle: Recovery Efforts; U.S. Charges 2 in Shuttle Debris Theft, Citing Need to 'Make an Example,'" *New York Times*, February 6, 2003; Matt Lait, "2 Texans Charged With Stealing Wreckage," *Los Angeles Times*, February 6, 2003, articles.latimes.com/2003/feb/06/nation/na-debris6; Jennifer Vose, "Punishments Vary for Debris Thieves," *Daily Sentinel* (Nacogdoches, TX), August 2, 2003; "Officer Cleared in Shuttle Debris Theft," *Chicago Tribune*, June 8, 2003, articles.chicagotribune.com/2003-06-08/news/0306080104_1_debris-stealing-shuttle; NASA Office of Inspector General press release, June 25, 2003.

25. Shafer and LeConey, "Legal Issues," 64.
26. ESRI, "Recovery Enhanced with GIS."
27. NASA's Lamar Russell took on the task of supporting the searches in the western states. He published a diary of his experiences: *The Silence and the Salvage* (Mustang, OK: Tate Publishing, 2013).
28. Cohrs, "Notes," 8.
29. Shafer and LeConey, "Legal Issues," 60.
30. Cohrs, "Notes," 8.
31. Gettleman, "Shuttle Debris Theft."
32. Cohrs, "Notes," 9.
33. NASA, "New Space Shuttle Columbia Images Released," news release 03-212, June 24, 2003. Searchers eventually recovered nearly ten hours of video and ninety-two photographs with in-cabin, Earth observation, and experiment-related imagery. Of the 337 videotapes aboard *Columbia*, twenty-eight were found with some recoverable footage. Only twenty-one rolls of film out of the 137 rolls of film aboard the ship were found with recoverable photographs.
34. Robert Crippen, remarks at the KSC Columbia Memorial Service, February 7, 2003, https://www.youtube.com/watch?v=ubYGeGU8jOo.
35. FEMA, "FEMA Updates Search, Find, and Secure Activities for Columbia Investigation," news release 3171-13, February 7, 2003.
36. Pat Oden email, February 8, 2003.
37. Interview with Larry Ostarly.
38. Interview with Scott Thurston.
39. "NASA Studies Possibility of Space Junk Role," *Florida Today*, February 6, 2003, 2S.
40. Mike Leinbach believes that perhaps the foam strike on the wing displaced an RCC panel on *Columbia*'s wing by compromising its support structure and pushing it back into the cavity behind the leading

edge. From there, it could have eventually broken off due to thermal expansion and contraction as the shuttle moved back and forth between orbital day and night. He personally still finds this theory more plausible than the idea that the foam actually punched a hole through an RCC panel.

41. "NASA: Search for Crucial Pieces Coming Up Short," *Florida Today*, February 6, 2003, 4S.
42. FEMA, "FEMA Updates Search, Find, and Secure Activities for Columbia Investigation," news release 3171-14, February 8, 2003.
43. Shafer and LeConey. "Legal Issues," 58–9. Phone calls and digital pictures emailed to the MIT closed most of these reports, as technicians were able to see that the material was clearly not related to the shuttle.
44. US Navy, *Salvage Report*, 1–8.
45. FEMA, "No Injuries Confirmed Because of Fallen Shuttle Materials; Citizens Urged to Avoid Contact With Unfamiliar Objects," news release 3171-15, February 8, 2003.
46. Interview with Ed Mango.
47. FEMA, "FEMA Responds to Offers of Donated Goods and Services for Columbia Emergency," news release 3171-16, February 9, 2003.
48. Interviews with Gerry Schumann, Don Eddings, Marsha Cooper.
49. Jeff Williams, interviewed by Connie Hodges.
50. Greg Cohrs added in an email to Jonathan Ward, "Shortly after the disaster, heavy rains repeatedly occurred, flooding the Attoyac River and Ayish Bayou, and all of the associated tributaries and watershed of Sam Rayburn and Toledo Bend Reservoirs. This flooding undoubtedly washed some debris away and covered other debris with water, probably hiding the items for a long time, if not forever."
51. US Navy, *Salvage Report*, 3–2.
52. US Navy, *Salvage Report*, 1–8.
53. William Harwood, "NASA Works to Eliminate Failure Scenarios," story written for CBS News *Space Place*, reprinted in *Spaceflight Now*, March 9, 2003, www.spaceflightnow.com/shuttle/sts107/030308scenarios/.

**Chapter 9: Walkers, Divers, and Spotters**

1. Interviews with Olen Bean, Greg Cohrs, and Mark Stanford. Cohrs added in an email to Jonathan Ward, "The Hemphill Management Team was filling individual resource orders for incident overhead

personnel beginning late in the first week, toward the weekend. Also, during the second week, the Southern Area (US Forest Service, Region 8) Type 1 Incident Management 'Blue' Team had assumed incident management of the Nacogdoches Camp activities. I'm guessing that this occurred by Wednesday, Day 12, because Marcus [Beard] and I drove over to Nacogdoches to share our experiences with the Blue Team personnel in the event that it might be helpful to them. I'm not sure when the other camps (Palestine and Corsicana) had IMTs assume command, except the Great Basin Type 1 IMT in-briefed with us in Hemphill on Sunday, February 16, and assumed command of the Hemphill Camp operations on Monday, February 17."

2. Keller, *USDA Forest Service Role*, 19.
3. A Type 1 "hotshot" crew is an elite, highly trained fire crew whose members always work together as a unit. A Type 2 crew is a regular fire crew made up of personnel who are available at a given time for an assignment.
4. NASA, *Report CB-QMS-024*, 9.
5. NASA, *Report CB-QMS-024*, 5.
6. James Hull email to all Texas Forest Service personnel, February 14, 2003.
7. Interviews with Dave King and Scott Wells.
8. Interviews with Mike Alexander.
9. "Shuttle Probe Exhausting," *Florida Today*, March 3, 2003, 3A.
10. "More Debris to Arrive at KSC Today," *Florida Today*, March 4, 2003, 2A.
11. Keller, *USDA Forest Service Role,* 44.
12. Texas Interagency Coordination Center, "Update—Space Shuttle Columbia Response," internal memo, February 14, 2003.
13. Southwest Texas Debriefing Team, "Space Shuttle Columbia Recovery Teams: A Grateful Nation Says 'Thank You,'" pamphlet distributed to incoming fire crews at Longview, TX, February 2003.
14. Keller, *USDA Forest Service Role*, 46.
15. Greg Cohrs email to Jonathan Ward.
16. Jerry Ross email to Jonathan Ward. The NASA astronauts who shared the "ground boss" role were Jim Halsell, Alan Poindexter, Bob Behnken, and Mike Foreman. Astronauts rotated during the recovery period to keep from getting worn out.

17. Greg Cohrs email to Jonathan Ward.
18. Interview with Gerry Schumann.
19. Shafer and LeConey. "Legal Issues," 79–82. NASA reviewed 153 property damage claims and provided compensation totaling $89,407. While many claims were legitimate, many were spurious. A typical example of the latter was one man who claimed that burning shuttle debris had set fire to his fishing pier, where in fact his barbeque grill had clearly caused the damage. The largest legal claim against NASA was by Spacehab, seeking $87.7 million in damages for loss of the research double module. Spacehab withdrew the claim in February 2007. "Spacehab Drops Columbia Lawsuit Against NASA; Says Efforts Better Spent Elsewhere," *Aero News Network*, February 22, 2007.
20. Cohrs, "Notes," 17.
21. Toward the end of the search period, the Forest Service set up a temporary helicopter base at Ennis, between Dallas and Corsicana.
22. Jerry Ross email to Jonathan Ward.
23. Interview with Boo Walker.
24. US Navy, *Salvage Report*, 2–4.
25. Jerry Ross email to Jonathan Ward. NASA's water search coordinators were astronauts Jim Reilly, Steve Bowen, and Keith Russell, assisted by a coast guard officer on loan to the Flight Crew Operations Directorate.
26. Texas Forest Service, "Situation Report Sunday Supplement: The Week in Review, April 13-20, 2003," internal memo.
27. Greg Cohrs email to Jonathan Ward.
28. US Navy, *Salvage Report*, sections 2–5.
29. Cohrs "Notes," 16.
30. "KSC Managers Visit East Texas Recovery Team," *Spaceport News* (Kennedy Space Center, FL), March 21, 2003, 1.
31. "Primary Search for Columbia Material Passes Halfway Mark," news release H03-117, March 25, 2003.
32. Interview with René Arriëns.
33. Gerry Schumann said that Pat Adkins was renowned for his ability to identify just about any piece of debris. The collection team occasionally amused themselves by tossing random pieces of metal, such as tractor parts, into the box. Adkins would pick them up, turn them over once in his hand, and ask, "Okay, who's the wiseass?"
34. Interview with René Arriëns.

35. Greg Cohrs emails to Jonathan Ward.
36. Interview with Jeremy Willoughby; Greg Cohrs emails to Jonathan Ward. While several other searchers from the Western United States claim to have discovered the OEX box, Cohrs's work plans from the day confirm that two Florida fire crews—Florida 3 and Florida 4—were working in the area the day the box was discovered.
37. Greg Cohrs email to Jonathan Ward.
38. William Harwood, "Recovered Data Tape in Relatively Good Condition," article for CBS News *Space Place*, reprinted in *Spaceflight Now*, March 24, 2003, www.spaceflightnow.com/shuttle/sts107/030324tape/.
39. Texas Forest Service, "Space Shuttle Columbia Recovery Efforts," internal memo, March 19, 2003.
40. Texas Forest Service, "Columbia Disaster Response 2003: NWCG Resources Mobilized Through Texas Interagency Coordination Center," March 24, 2003.

**Chapter 10: Their Mission Became Our Mission**

1. Interview with Jim Furr.
2. Interview with Pat Adkins.
3. Kenneth Ward email to Jonathan Ward.
4. Interview with Boo Walker.
5. Christopher Freeze, "The 'Columbia' Debris Recovery Helo Crash, March 27, 2003," www.check-six.com/Crash_Sites/STS107-N175PA.htm.
6. Greg Cohrs email to Jonathan Ward.
7. Interview with Ed Mango.
8. Freeze, "Debris Recovery Helo Crash."
9. Interview with Ed Mango.
10. Freeze, "Debris Recovery Helo Crash."
11. Interview with Jeff Angermeier.
12. Interview with Ed Mango.
13. FEMA, "Space Shuttle Columbia; Emergency and Related Determinations," news releases FEMA-3171-EM-TX and FEMA-3172-EM-LA and situation report 59, April 7, 2003.
14. NASA, "Columbia Recovery Visit to Nacogdoches Incident Command Site and Lufkin, Agenda for STS-114 Crew," NASA Public Affairs Office (Johnson Space Center) internal memo, April 10, 2003.

15. Stacy Faison, "Shuttle Crew Visits East Texas," *Lufkin Daily News*, April 11, 2003.
16. NASA, "Columbia Recovery Agenda for Family Visit," internal memo, April 23, 2003.
17. Interview with Brent Jett.
18. US Navy, *Salvage Report*, 5-2.
19. Cohrs, "Notes," 19.
20. NASA, *Report CB-QMS-024*, 9.
21. USDA Forest Service, "Fire and Aviation Management Briefing Paper, Columbia Support, Interagency Support to Space Shuttle Columbia Recovery Effort," USDA Forest Service, Washington, DC, May 2, 2003.
22. Jan Amen email, April 30, 2003.
23. "Columbia Shuttle Recovery Appreciation Program" (Lufkin, TX: April 29, 2003), program agenda.
24. Interview with Boo Walker.
25. US Navy, *Salvage Report*, 5–6, B-1.
26. FEMA, "Recap of the Search for Columbia Shuttle Material," news release 3171-EM NR071, May 5, 2003.
27. Mark Stanford, "STS-107 Space Shuttle Columbia Recovery Operation, February 1–May 10, 2003," PowerPoint presentation (undated); David King and Scott "Doc" Horowitz, "Space Shuttle Columbia Recovery Operation," (Washington, DC: March 8, 2013), presentation to Space Policy Institute Symposium. https://www.c-span.org/video/?311395-3/space-shuttle-columbia-recovery-operation.

## Chapter 11: Reconstructing *Columbia*

1. Interview with Steve Altemus.
2. Interview with Steve Altemus.
3. NASA, untitled video on lessons learned in *Columbia* reconstruction with Steve Altemus and Pam Melroy, July 2003 (unreleased).
4. Pam Melroy, NASA reconstruction video.
5. Michelle La Vone, "The Space Shuttle Challenger Disaster," *Space Safety Magazine*, January 28, 2016, www.spacesafetymagazine.com/space-disasters/challenger-disaster/.
6. Robert Pearlman, "Smithsonian Considering Display of Fallen Shuttles Challenger and Columbia Debris," *collectSPACE.com*, January 31, 2011, www.collectspace.com/news/news-013111a.html.

7. Interview with John Biegert.
8. Interview with Jim Comer.
9. Interviews with Jim Comer and Jon Cowart.
10. Pam Melroy, in NASA reconstruction video.
11. "Schirra, Lovell Cheer KSC Workers," *Spaceport News*, March 21, 2003, 7.
12. Interview with Pat Adkins.
13. Interview with Jim Comer.
14. Interviews with Steve Altemus and Pam Melroy.
15. "Panel Confident of Finding Cause," *Florida Today*, March 19, 2003, 2A.
16. NASA, *Columbia Accident Investigation Board Report, Vol. 1* (Washington, DC, August 2003), 75.
17. Interview with Marty McLellan.
18. Shafer and LeConey. "Legal Issues," 56–7.
19. "Student Science Project Survived Shuttle Disaster," CNN, May 24, 2003. Quoted in Liston, *Chronology of KSC for 2003*, 111.
20. NASA reconstruction video.
21. *CAIB Report*, 75.
22. *CAIB Report*, 76.
23. *CAIB Report*, 75.
24. NASA reconstruction video.

**Chapter 12: Healing and Closure**

1. Sean O'Keefe email to Mike Leinbach and Jonathan Ward.
2. Interviews with Ann Micklos, Pam Melroy, and John Biegert. Biegert said that many of the digital timers used by the crew survived reentry and were recovered in working order, although their displays were fogged over.
3. Interview with Jim Wetherbee.
4. Jeff Williams, interviewed by Connie Hodges.
5. *CAIB Report*, 61.
6. Jim Comer noted that *Enterprise* was at the time being prepared for exhibit in the Udvar-Hazy Center of the Smithsonian's National Air and Space Museum. Pam Melroy and Comer traveled to Washington and negotiated with museum director Gen. J. R. "Jack" Dailey for a loan of the leading edge panels and the landing gear door. As a side note, *Enterprise* did not have thermal tiles, since it was not intended

to fly in space. NASA glued tiles to *Enterprise*'s landing gear door to simulate an operational shuttle for these tests.

7. Interview with Steve Altemus.
8. Interview with Robert Hanley.
9. "Debris Reconstruction Hangar Walk-through Days Scheduled," *Spaceport News*, June 27, 2003, 2.
10. "Texas Family Recalls Recovery Contributions," *Spaceport News*, September 5, 2003, 4.

## Chapter 13: Preserving and Learning from *Columbia*

1. Robert Pearlman, "Smithsonian Considering Display."
2. Interview with Mike Ciannilli.
3. Interview with Scott Thurston.
4. "Storage of Columbia Debris to be Determined," *Spaceport News*, July 11, 2003, 2.
5. Interview with Scott Thurston.
6. "Columbia Tank Found on Lakebed," NASA online article, August 3, 2011, https://www.nasa.gov/mission_pages/shuttle/main/columbia-tankfound.html.
7. "Columbia Debris Finds Final Home in VAB," *Spaceport News*, October 31, 2003, 2.
8. "VAB 16th Floor A Tower Is Columbia's Arlington," *Spaceport News*, February 13, 2004, 5.
9. STS-121 mission summary, https://www.nasa.gov/mission_pages/shuttle/shuttlemissions/archives/sts-121.html.
10. Interviews with Mike Leinbach and Mike Ciannilli.
11. Interviews with Jim Comer and Steve Altemus.
12. NASA, *Crew Survival Investigation Report*, 4–5.
13. Pam Melroy email to Jonathan Ward; NASA, *Crew Survival Investigation Report* 3-69–3-70.

## Chapter 14: The Beginning of the End

1. NASA, "Expedition 6 Crew Returns Home," May 3, 2003, https://www.nasa.gov/missions/shuttle/soyuz_landing_update.html.
2. Jim Banke, "NASA's O'Keefe Promises Study of Safety Reporting System," Space.com, May 22, 2003, quoted in Liston, *KSC Chronology of KSC for 2003*, 109–10.
3. *CAIB Report*, 118.

4. "Inquiry costs taxpayers $454 million," *Florida Today*, August 26, 2003, 1A, 5A.
5. Wayne Hale notes to Mike Leinbach.
6. Hubble orbits at 28.5° inclination and 335 miles altitude; the ISS orbit is 51° inclination and 250 miles altitude.
7. Space Telescope Science Institute, "Servicing Mission 4 Cancelled," status report January 15, 2004, www.spaceref.com/news/viewsr.html?pid=11615.
8. NASA, "NASA Approves Mission and Names Crew for Return to Hubble," news release 06-343, October 31, 2006.
9. "Investigation Could Reshape Space Agency," *Florida Today*, March 3, 2003, 3A, quoted in Liston, *KSC Chronology of KSC for 2003*, 52.
10. "Designer: Suspend Human Space Program," *Florida Today*, May 16, 2003, 11A, quoted in Liston, *KSC Chronology of KSC for 2003*, 109–10.
11. The International Space Station Intergovernmental Agreement is an international treaty signed on January 29, 1998, by fifteen governments for "a long term international co-operative frame-work on the basis of genuine partnership, for the detailed design, development, operation, and utilisation of a permanently inhabited civil Space Station for peaceful purposes, in accordance with international law."
12. "Return ASAP," *Aviation Week & Space Technology*, September 29, 2003, 21.
13. Andrew Thomas, "Exploration—To the Fire of the Human Spirit: A Tribute to Fallen Astronauts and Cosmonauts," August 4, 2005, https://www.nasa.gov/returntoflight/crew/sts114_exp11_tribute.html.
14. FEMA's Scott Wells and Mark Stanford of the Texas Forest Service, among many others who participated in the *Columbia* recovery, were also involved in responding to the aftermath of Hurricane Katrina. While lessons learned from the *Columbia* operation actually improved the coordination among many of the agencies during the Katrina response, leadership breakdowns in other areas is a topic best discussed elsewhere.
15. Warren Leary, "Cracks Found in Protective Foam on an Unused Shuttle Fuel Tank," *New York Times*, November 23, 2005.
16. Wayne Hale, "How We Nearly Lost *Discovery*," *Wayne Hale's Blog*, April 18, 2012, https://waynehale.wordpress.com/2012/04/18/how-we-nearly-lost-discovery/.

17. "Bird Droppings Survive Space Launch," *Washington Post*, July 5, 2006, http://www.washingtonpost.com/wp-dyn/content/article/2006/07/05/AR2006070501242.html.
18. Justin Ray, "NASA Space Shuttle Launch Director Joins Commercial Rocket Company," *Space.com*, January 24, 2012, www.space.com/14333-shuttle-launch-director-leinbach-joins-ula.html.

**Chapter 15: Celebrating 25,000 Heroes**

1. National Park Service, *Space Shuttle Columbia Memorial Special Resource Study*, National Park Service Intermountain Region (Denver, CO: October 2014), 86.
2. National Park Service, *Columbia Memorial Resource Study*, 9.
3. A few other pieces of *Columbia* debris are in display cases in administrative buildings at NASA Centers. For example, the crew hatch window is in the KSC Headquarters Building, and the OEX recorder is at JSC. These locations are not typically accessible to the general public, however. Mike Ciannilli was instrumental in obtaining approval to exhibit the hatch and hatch cover from the Apollo 1 spacecraft, which can be seen at KSC's Apollo/Saturn V Visitors Center.
4. Cohrs, "Notes," 2.

# INDEX

# ABOUT THE AUTHORS

**Michael D. Leinbach** was the final Shuttle Launch Director at NASA's John F. Kennedy Space Center (KSC), Florida. He was responsible for overall Shuttle launch countdown policy, planning, and execution activities.

Leinbach joined NASA in 1984 as a structural engineer in the Design Engineering Directorate. He served as a lead design engineer for a variety of launchpad systems including the Orbiter Weather Protection and Emergency Egress Slidewire systems. In 1988, he became a NASA Test Director (NTD) in the Shuttle Management and Operations Directorate (later, the Shuttle Processing Directorate). As an NTD, he was responsible for directing all daily operations at Launch Complex 39. Concurrently, he was chairman of the Emergency Egress and Rescue Working Group. In 1991, he was named Shuttle Test Director, conducting the terminal countdown and launch of seventeen Shuttle missions and was responsible for all prelaunch planning activities involving the Shuttle launch countdown.

From January 1998 to May 2000, Leinbach served as the deputy director of the Space Station Hardware Integration Office, where he was responsible for all International Space Station (ISS) component processing at KSC and contractor manufacturing locations. Leinbach also oversaw the development and execution of the Multi-Element Integrated Test Program, which verified the functionality and operability of the first phase of the ISS program in a configuration, on the ground, as close to the on-orbit final assembly as possible.

Leinbach was tapped to serve as Assistant Launch Director in May 2000 and was named Launch Director in August 2000. He led the Launch Team for all Shuttle missions from then to the end of the program in 2011, serving as the person to give the final "Go" for launch. He also served as the senior operations expert for NASA for all Shuttle flight elements and ground support equipment processing issues.

Immediately following the *Columbia* accident in February 2003, Leinbach led the initial debris recovery effort in Texas and Louisiana. Shortly thereafter, he was named to lead the *Columbia* reconstruction team chartered to determine the cause of the accident based solely on the debris collected and reassembled at KSC. He was also the driving force behind the *Columbia* preservation team and development of the plan to lend debris to academia for study, with the goal of developing better and safer spacecraft in the future.

In November 2004, Leinbach was awarded the prestigious 2004 Presidential Rank Award. He has received numerous group achievement and performance awards, including NASA's Exceptional Service Medal in 1993 for his leadership in planning and conducting Shuttle launch countdowns and NASA's Medal for Outstanding Leadership in May 2003 for significant contributions to the Space Shuttle Program.

Born in Reading, Pennsylvania, he graduated in 1971 from Yorktown High School in Arlington, Virginia. He received a BS in Architecture in 1976 and a Master of Engineering in Civil Engineering with emphasis in structural dynamics in 1981 from the University of Virginia in Charlottesville.

Leinbach remains active in public outreach and education at Kennedy Space Center. He leads the monthly "Launch Director Tour" from the KSC Visitors Center complex, in which he explains the risks and rewards of human spaceflight and takes his groups to some of the key facilities at KSC. Leinbach also mentors new and mid-career employees through various leadership and educational forums.

Leinbach retired from NASA in 2011. He and his wife Charlotte reside in Scottsmoor, Florida.

**Jonathan H. Ward** spent several years of his childhood in Japan and considers the Virginia suburbs of Washington, DC, to be his hometown. Although he has a wide variety of interests and has worked in many fields, space exploration is his lifelong passion. His joy of bringing the space program to life for the general public began in high school, when he served as a volunteer tour guide at the Smithsonian's National Air and Space Museum during the Apollo 15 and Apollo 16 missions. He continues his public outreach today, as a Solar System Ambassador for NASA's Jet Propulsion Laboratory, as a frequent speaker on space exploration topics to interest groups and at regional conferences, and as an author of books on space history. He is also a regular contributor to online space exploration forums.

Ward brings a unique perspective to his writing that marries a systems view of the topic, fascination with the technology, passion for space exploration, and deep respect for the people who make it all happen.

After studying physics at Carnegie-Mellon University, he transferred to Virginia Commonwealth University, from which he graduated *summa cum laude* in 1978 with a BS in Psychology. He received a Master of Science in Systems Management degree from the University of Denver in 1992.

Ward is professionally certified as an executive coach by the International Coach Federation and serves on the adjunct staff at the Center for Creative Leadership. His professional experience includes extensive work as an organizational development and leadership consultant, both as an employee of some of America's largest companies and as an external consultant. He worked for several years with Boeing on the Space Station Freedom program. His varied other roles have included strategic systems planning at Freddie Mac and leading Capital One's global effectiveness functions.

Ward's book *Rocket Ranch* was the bestselling engineering title for technical publisher Springer Books in 2015—quite an accomplishment for a non-engineer author. His books on the Apollo/Saturn program at Kennedy Space Center have been praised as "the perfect balance between presenting the technical aspects of launch preparations and the personal side of what it must have been like to be a part of the Apollo workforce during this incredible time in the history of manned space flight. The overall flow of these two books really makes the reader feel like they have gone back in time and were given their own personal VIP tour of the Kennedy Space Center."

Roger Launius of the Smithsonian's National Air and Space Museum said, "Jonathan Ward's *Rocket Ranch* is an enjoyable overview of the Kennedy Space Center during the 1960s . . . For those who are interested in Apollo era technology infrastructure at the Kennedy Space Center, this is the book for you."

*Quest: The History of Spaceflight Quarterly* said, "*Countdown to a Moon Launch* . . . merits a space on the bookshelf of any Apollo aficionado. Let's hope that Ward finds time to write more such books."

Ward and his wife Jane now reside in Greensboro, North Carolina.